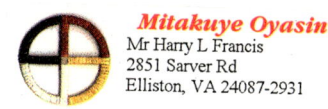
Mitakuye Oyasin
Mr Harry L Francis
2851 Sarver Rd
Elliston, VA 24087-2931

To my good friend Harry

All my best

HANDBOOK FOR STABILIZATION OF PAVEMENT SUBGRADES AND BASE COURSES WITH LIME

PREPARED FOR:

    APG LIME COMPANY
    AUSTIN WHITE LIME COMPANY
    CHEMICAL LIME GROUP
    REDLAND STONE PRODUCTS COMPANY
    TEXAS LIME COMPANY

BY:

    DALLAS N. LITTLE
    KELLEHER PROFESSOR IN ENGINEERING
    TEXAS A&M UNIVERSITY
    AND ASSISTANT DIRECTOR,
    TEXAS TRANSPORTATION INSTITUTE

Edited by:
Harry L Francis
Technical Mgr.

SPONSORED BY THE NATIONAL LIME ASSOCIATION

KENDALL/HUNT PUBLISHING COMPANY
4050 Westmark Drive   Dubuque, Iowa 52002

Copyright © 1995 by National Lime Association

Library of Congress Catalog Card Number: 94-78821

ISBN 0-8403-9632-5

All rights reserved. No part of this publication may be reproduced, stored in a retrieval system, or transmitted, in any form or by any means, electronic, mechanical, photocopying, recording, or otherwise, without the prior written permission of the copyright owner.

Printed in the United States of America
10 9 8 7 6 5 4 3 2 1

TABLE OF CONTENTS

Page

CHAPTER 1   INTRODUCTION TO THIS HANDBOOK 1
1.01   Purpose of the Handbook 1
1.02   Definition of General Terms 1
      Subgrade Stabilization 1
      Base Stabilization 2
      Lime Modification 2
      The Construction Procedure 3
1.03   Organization of the Handbook 3
1.04   Effective Use of the Handbook 4
1.05   Example Use of the Handbook 5
1.06   References 6

CHAPTER 2   INTRODUCTION TO LIME 7
2.01   Definition of Terms 7
2.02   Forms of Lime Generally Available for Stabilization 10
2.03   Important Physical Properties of Quicklimes and Hydrated Limes 10
      Specific Gravity and Bulk Density 11
      Particle Size 12
      Heat of Formation 12
      Solubility of Hydrated Lime 12
2.04   Important Chemical Properties of Quick and Hydrated Limes .. 13
      pH of Lime-Water Solutions 14
      Rate of Solution 14
      Reaction of Lime With Carbon Dioxide 14
      Reaction With Silica and Alumina 15
      Environmental Effects 15
2.05   Production of Lime 16
2.06   Lime Industry 18
2.07   References 18

CHAPTER 3   ROLE OF STABILIZED LAYERS IN THE PAVEMENT STRUCTURE 19
3.01   Role of the Surface Course 19
3.02   Role of Base and Subbase Courses 19
3.03   Role of the Base and Subbase in Minimizing Pavement Distress in Flexible Pavements 21
3.04   Important Characteristics of Lime-stabilized Bases, Subbases and Subgrades to Meet Structural Pavement Demands 22
3.05   Use of Lime in Upgrading Subgrade Soils and Aggregate Bases 25
3.06   References 27

CHAPTER 4   MECHANISMS OF LIME-SOIL STABILIZATION 29
4.01   General 29
4.02   Need for Stabilization 29
4.03   Nature of Soil (Soil Mineralogy) 30
      The Clay Minerals 32
      Clay-Water System 35
4.04   The Dramatic Change: The Lime-Clay System 37
      A Stable Water Layer 37
      A New Texture 38
      Immediate Strength Improvement 39
      Long Term Strength 41
4.05   Evidence Supporting the Lime-Soil Reaction 43
4.06   Suitable Soils for Lime Stabilization 49
4.07   Factors Which Influence Lime-Soil Interactions 49
      Organic Carbon 50
      The Effect of Sulfates on Lime-Soil Interactions 51
      Clay Content 52
      Nature of Clay (Mineralogy) 52
      Weathering 53
      Pedology 53
      Geological and Climatic Effects 54

4.08	Important Sources of Information Concerning Soil-Lime Reactivity Potential 55
4.09	References 57

CHAPTER 5 MIXTURE DESIGN 59

5.01	Objective of Mixture Design 59
5.02	Mixture Design Criteria 59
	Modification 59
	Stabilization 60
5.03	Current Mixture Design Procedures 60
	Thompson Procedure 61
	Treatment Level 61
	Mixture Design Protocol 61
	Mixture Preparation 61
	Density Control 61
	Curing Conditions 62
	Testing Procedures 62
	Mixture Design Criteria 63
	Proposed Mixture Design Process 64
	Eades and Grim Procedure 64
	Texas Procedure 67
5.04	Accelerated Curing Cautions 68
5.05	Summary 69

APPENDIX 5.01 COMPRESSION STRENGTH OF MOLDED SOIL-LIME CYLINDERS 71

A. Purpose 71
B. Scope 71
C. Applicable Documents 71
D. Apparatus 71
E. Preparation of Soil-Lime Mixture 72
F. Molding Specimens 72
G. Compression Test 73

H. Calculation 73
I. Report 73
5.06 References 74

CHAPTER 6   ENGINEERING PROPERTIES OF LIME-STABILIZED SOILS AND AGGREGATES 75
6.01   Properties and Characteristics of Soil-Lime Mixtures 75
     Uncured Mixtures 76
        Plasticity 76
        Moisture Density Relationships 77
        Swell Potential 78
        Textural Changes 81
        Strength and Deformation Properties 81
        Summary 84
     Cured Mixtures 84
        Strength Properties 84
        CBR 86
        Poisson's Ratio 92
     Permeability 93
     Shrinkage of Lime-Soil Mixtures 93
     Durability of Lime-Soil Mixtures 94
     Compression 96
     Variability of Lime-Soil Mixtures 97
6.02   Engineering Properties of Lime-Soil Mixtures Under Field Conditions (in situ) and Under Simulated Field Loading Conditions (lab) 98
     General 98
     In Situ Moduli of Aggregate Base Course Stabilized With Low Percentages of Lime 102
     In Situ Moduli and Strength Properties of Lime Stabilized Subgrades 105
     Application of In Situ Moduli to Pavement Design 110
6.03   Long-term Strength of Lime Stabilized Mixtures 114
6.04   Longevity of Lime Treated Soils 121

	Durability of Lime Stabilization Construction Projects 121
	Leaching Effects on Soil-Lime Mixtures 121
	McCallister-Petry Study 122
6.05	Strength Enhancement of Lime Soil and Lime Aggregate Mixtures through the Addition of Fly Ash and Lime Fly Ash (LFA) 123
	General 123
	Available Calcium 124
	Suitable Soils and Aggregates 124
	Engineering Properties 125
6.06	References 125

CHAPTER 7	THICKNESS DESIGN CONSIDERATIONS 129
7.01	Pavement Design Strategies 129
	Empirical Methods and Limiting Shear Failure Methods 129
	Limiting Deflection Methods 131
	Regression Methods Based on Pavement Performance or Road Tests 131
	Mechanistic-Empirical Methods 134
7.02	Example of the Influence of Lime Treated Subgrade Layer in a Flexible Pavement System 135
7.03	Example of the Role of Lime Treated Subgrade in a Rigid Pavement Design 139
7.04	References 141

CHAPTER 8	LIFE CYCLE COSTING 143
8.01	Pavement Type Selection 143
8.02	Life Cycle Cost Analysis 144
	Costs Associated With Pavement Rehabilitation 145
	Discount Rate 146
	Constant Dollar Studies 146
	Current Dollar Studies 146
	Discussion 147

Recommendation 147
Analysis Life 149
Salvage Value 150
Life of Rehabilitation Alternatives 150
8.03 Cost Data 151
8.04 Analysis Procedure 152
8.05 Sensitivity Analysis 153
8.06 Example Problem 153
8.07 References 159

CHAPTER 9 CONSTRUCTION OF LIME STABILIZED BASES AND SUBBASES 161

9.01 Lime Treatment Methods 161
In-Place Mixing 161
Plant Mixing 162
Pressure Injection 162
9.02 Steps in the Construction Process 164
Soil Preparation 164
Lime Application 165
Dry Hydrated Lime 165
Dry Quicklime 166
Slurry Method 168
Advantages and Disadvantages of Different Lime
Applications 172
Dry Hydrated Lime 172
Dry Quicklime 172
Slurry Lime 173
Pulverization and Mixing 173
One-Stage Mixing 173
Two-Stage Mixing 174
Blade Mixing 174
Rotary Mixing 177
Central Mixing and Upgrading
Marginal Aggregates with Lime 177

		Pulverization and Mixing Requirements 179
		Compaction 179
		Curing 180
	9.03	Construction Considerations for Lime and Fly Ash Applications 183
	9.04	Measurement and Payment 185
	9.05	References 186

CHAPTER 10 QUALITY CONTROL AND GUIDE SPECIFICATIONS FOR LIME-TREATED LAYERS 187

10.01 Field Quality Control Considerations 187
      Lime Spread Rate 187
      Pulverization 187
      Mixing Efficiency 187
      Depth of Lime Treatment 188
      Lime Content 188
      Moisture Content 188
      Density 188
      Slurry Composition 189
      Weather Limitations 189
10.02 Guide Specifications for Lime-Treated Subgrades 189
      1.0. General 189
          1.1 General Requirements 189
          1.2 Submittals 189
          1.3 Delivery and Storage 190
          1.4 Weather Limitations 190
      2.0 Materials 190
          2.1 Lime 190
          2.2 Soil 192
          2.3 Water 192
          2.4 Bituminous Curing Seal 192
      3.0 Execution 192
          3.1 Sequence of Construction Operations 192
          3.2 Site Preparation 192

	3.3 Lime Treatment 193
	3.4 Traffic Control, Curing, Maintenance, and Drainage Protection 195
	3.5 Equipment Limitations 195
	3.6 Safety Requirements 196
	3.7 General 196

10.03 Guide Specifications for Lime-Treated Base Courses 197
    1.0 Scarification and Pulverization 197
    2.0 Lime Spreading 197
    3.0 Mixing and Watering 198
    4.0 Compaction 199
    5.0 Curing 199

10.04 Central Mixed Stabilization Procedures 199
    1.0 Central Mixing 199
    2.0 Placing Material 199
    3.0 Compaction 199
    4.0 Curing 199

10.05 Lime Modification 200
    1.0 Base Material 200
    2.0 Subgrade Materials 200

10.06 Applicable Publications to Lime Stabilization of Subgrades, Subbases or Bases 200
    American Association of State Highway and Transportation Officials (AASHTO) Publications 200
    American Society for Testing and Materials (ASTM) Publication, Latest Edition 201
    Other Important Specifications and Publications 202

10.07 Precautions 202
    Amount of Lime 202
    Amount of Water 203
    Compaction 203
    Curing 203
    High Sulfate Soils 204

CHAPTER 11 CONSIDERATIONS FOR USING LIME STABILIZATION IN PAVEMENT RECYCLING 205

11.01 Recycling Alternatives 205
    In-Place Recycling 206
    Central Plant Recycling 206
11.02 Analysis and Design Steps Required for In-Place Recycling 206
    Equipment and Methods 206
    Application of In-Place Recycling Techniques 207
    Construction Procedures 207
    Mixture Design 208
    Structural Design 208
    Energy and Cost Considerations 208
    Construction Specifications and Quality Control 208
11.03 Case History of the Use of Lime in In-Place Recycling 208
    Recycling of City Streets in Waco, Texas 208
        Overview of Project 208
        Planning and Construction Process 208
        Cost Savings 209
        Project Success as Determined From Visual
        and Serviceability Data 209
    Yolo California Recycling Project 210
11.04 References 210

CHAPTER 12 LIME SLURRY PRESSURE INJECTION 211

12.01 Introduction 211
12.02 Lime Slurry Injection Process 211
12.03 Injection Materials 213
12.04 LSPI Mechanisms 213
    Stabilizing Effect of Lime Seams 214
    Preswelling 214
    Translocation 214
    Supernate Penetration 215
12.05 Candidate Soils for LSPI 215

        Glaze Stabilized Compression Tests 215
        Seam Stabilized Compression Test 216
        Glaze Stabilized Consolidation Test 217
        Seam Stabilized Swell Test 217
        Material Test 217
        Field Pump Tests 218
        Surcharge Tests 218
        The Decision Process 218
12.06 Safety Precautions 218
12.07 Conclusions 218
12.08 References 219

LIST OF FIGURES

Figure   *Page*
  2.1. Rotary Kilns are Used in the Production of Lime 16
  2.2. Lime Has Many Uses in Industry (After NLA
       Bulletin 214, 1992) 17
  3.1. Subbase Below PCC Pavement Provides Uniform Support
       and Aids in Load Transfer 20
  3.2. Critical Parameters in a Flexible Pavement Include Flexural
       Tensile Strain in the HMAC, Shearing Stresses within the HMAC
       and Vertical Compressive Stresses and Strains at the Top of the
       Subgrade 21
  3.3. Tensile Strains and Stresses in the HMAC are Reduced as a
       Result of Improved Support of the HMAC and Aggregate Base
       Course by the Stabilized Subgrade Layer 22
  3.4. The Distribution of Vertical Compressive Stresses and Strains
       are Controlled by the Stiffnesses or Moduli of the Base and
       Subbase Layers 23
  3.5. The Resilient Modulus Test is a Primary Method of
       Characterizing Pavement Materials for Use in Pavement
       Layers. The Laboratory Test (a) Attempts to Simulate Field
       Loading Conditions (b) 24

3.6. Stiff (High Modulus) Stabilized Subbases or Bases Can Develop High Flexural Stresses Which Should Be Considered in Layer Thickness Design 26
4.1. Two of the Most Common Building Blocks of Soils are the Silica Tetrahedron (a) and the Aluminum Octahedron (b) 31
4.2. Silica Tetrahedra in a Lattice Arrangement Produce Very Stable Minerals Such as Quartz Sand. (After Moffatt et al., 1965) 31
4.3. Tetrahedra and Octahedra are Units Which Can Also Be Arranged in Sheets with Large Surface Areas 33
4.4. Basic Units of the 1:1 Mineral Kaolinite (a) Are Linked with Relatively Strong Hydrogen Bonds Which Retain a High Degree of Moisture Stability Among Layers While the Basic Units of the 2:1 Smectite Mineral (b) Are Linked by Weak Cation Attraction. The Efficiency of This Linkage is a Function of the Type and Concentration of the Available Cations 34
4.5. Cations and Water (a Dipolar Molecule) Are Attracted to the Negatively Charged Clay Surface to Satisfy the Charge Potential. This Results in (a) Adsorbed Cations and Water Molecules and (b) a Diffused Layer of Cations Due to Their Thermal Activity and the Infusion of Water Toward the Clay Surface Because of the High Electrolyte Concentration. (After Mitchell, 1976) 36
4.6. The Reason for the Textural Change is Due to the Phenomenon of Cation Exchange Followed by Flocculation and Agglomeration. (a) Illustrates Low Strength Clay Soil Where Particles Are Separated by Large Water Layers. The Addition of Lime (Calcium) Shrinks the Water Layer (b) Allowing the Plate-Like Particles to Flocculate 38
4.7. Lime Treatment Provides Immediate Strength Gains as Illustrated by These CBR Data as a Function of Moisture Content for a CL Soil. (After Thompson, 1970) 40
4.8. The Effect of a High pH System is to Release Silica and Alumina From the Clay Surface. (After Keller, 1964) 42

4.9. The Quantity of Lime Required to Produce the Pozzolanic Reaction (as Reflected by Compressive Strength) Varies With the Type and Mineralogy of the Soil Being Stabilized (After Eades and Grim, 1960) 45

4.10. X-Ray Diffraction (XRD) Spectra Can Be Used to Prove the Reaction Occurring Between the Lime and the Clay Surface. In (a) the XRD Spectra of the Natural Soil Produces Intense Smectite Peaks While the Peak Essentially Diminishes Upon Lime Treatment (b) 46

4.11. Scanning Electron Micrographs (SEM's) Provide Visual Evidence of the Development of Pozzolanic Crystals in This Denver, Colorado Clay (b) the Natural Soil Without Lime is Shown in (a) 47

4.12. The Crystals Produced in Soils Vary Depending on Soil Mineralogy and Time of Curing. This Arlington, Texas, Soil (a) Demonstrates Considerable Pozzolanic Product Development Upon Addition of Lime (b) 48

4.13. A Good Estimate of the Appropriate Stabilizer For a Certain Soil Can Be Determined Based on Simple and Easy to Measure Soil Index Properties Such as Minus 200 Content and PI (After Currin et al., 1976) 50

4.14. Illustration of Information Available in County Soil Report (Travis County, Texas) 56

5.1. The Thompson Mixture Design Flow Chart is Based On Soil-Lime Reactivity (After Little et al. (1987)) 65

5.2. The Eades and Grim pH Test Is An Excellent Indicator of Optimum Lime Content and Should Be Used As Part of a Complete Mixture Design Procedure To Insure Optimization of Pozzolanic Reactivity 67

5.3. The Texas Department of Transportation Selects Optimum Lime Content for Strength Testing Based on the Soil Index Properties of Soil Binder (Minus No. 40 Sieve Size) and PI 69

6.1. Different Percentages of Lime are Required to Reduce Plasticity to Desired Levels for Different Soils. This Reaction is Immediate in That it Does Not Require Long Curing Times. (After Holtz, 1969) 76

6.2	Even Soils Which Require Long-Term Curing (At Least 28-Days) To Develop Significant Strength Gains Demonstrate Immediate PI Reduction As Demonstrated By This Beaumont Clay	77
6.3	The Shift In Density And Optimum Moisture Content For Achieving Maximum Density Is Evidence Of The Physical Changes That Occur (Immediately) During Lime Treatment. (After Terrel et al., (1979))	78
6.4.	Plasticity Index Can Be Used To Predict Swelling As Demonstrated By Seed et al. (1962)	79
6.5.	Swell Pressure As A Function of Lime Content and Period of Curing For Irbid, Jordan, Clay (After Basma and Tuncer, 1991)	81
6.6	Effect Of Lime And Curing Time On (a) Plasticity and (b) The Clay-Size Fraction Of Irbid, Jordan Clay. (After Basma And Tuncer, 1991)	82
6.7.	Immediate Effects Of Lime Treatment Can Be Substantial And Are Demonstrated By This 10-Fold Improvement In Resilient Modulus. (After Little et al. (1987))	83
6.8.	Although Accelerated Curing Procedures Can Be Used To Approximate Long-Term Strength Gain, Lime-Soil Pozzolanic Reactions Occur Over Time. The Long-Term Beneficial Pozzolanic Effects Should Be Accounted For In Design	86
6.9.	Effect of Lime Content and Curing Time on Undrained Cohesion (a) and Undrained Angle of Internal Friction (b) for Jordanian Soil (After Tuncer and Basma, 1991)	88
6.10.	Texas Department of Transportation Uses a Triaxial Test to Evaluate the Acceptability of Granular (Unbound and Stabilized) Bases and Subbases	89
6.11.	Lime Stabilization of a Colorado River Gravel, Rich in Clay Binder, Improves the Texas Triaxial Classification	89
6.12.	As the Strength of a Soil-Lime Mixture Increases with Curing, the Stiffness of the Mixture Does Also (After Suddath and Thompson, 1975)	91
6.13.	The Stress-Strain Curve of This Burleson, Texas, Clay is Substantially Changed by Lime Stabilization. The Slope of the Initial Portion of the Curve Represents the Stiffness or Modulus of the Mixture (After Shrinivas, 1992)	92

6.14. If Adequate Lime is Available, Pozzolanic Reaction Will Continue to Occur Under Favorable Curing Conditions Resulting in Autogenous Healing During Favorable Seasons (After Thompson and Dempsey, 1969) 95

6.15. Effect of Lime and Curing Time on Rate of Compression For Jordanian Soil (After Basma and Tuncer, 1991) 97

6.16. The Resilient Modulus of Granular Soils and Aggregates is a Function of the Stress State and Moisture Content and is Typically Presented as a Function of the Bulk Stress 98

6.17. The Resilient Modulus of Fine-Grained Soils is a Function of the Stress State and Moisture Content and is Typically Presented as a Function of Deviator Stress 99

6.18. Illustration of Seasonal Variability of Resilient Modulus of a Fine-Grained, Cohesive Soil 100

6.19. AASHTO Has Divided the United States Into Six Climatic Regions for Pavement Design and Analysis (After AASHTO Pavement Design Guide, 1986) 101

6.20. FWD Deflection Basins for Six Phoenix, Arizona, Pavements Were Substantially Influenced Based on Whether or Not the Aggregate Base Course (ABC) was Stabilized with a Low Percentage of Lime (After Little, 1990) 103

6.21. Variation in CBR, as Approximated by DCP, with Depth for Lime-Stabilized and Natural Subgrade for FM 2818 Near Bryan, Texas (After Scullion and Little, 1993) 109

6.22. Variation in CBR with Depth for Flexible Hard, Lime-Stabilized Subgrade and Natural Subgrade for FM 3478 (After Scullion and Little, 1993) 110

6.23. Structural Layer Coefficient, a_2, Was Determined by Thompson As a Function of Compressive Strength for Lime Stabilized Layers (After Thompson, 1970) 112

6.24. Pavement Deflections Were Measured with a Benkleman Beam. Subgrades Contained Various Levels of Lime Treatment in a Nebraska Study Over a Three Year Period (After Lund and Ramsey, 1959) 119

6.25. Pavement Deflections Were Measured with a Benkleman Beam. Bases Contained Various Levels of Lime Treatment in a Nebraska Study Over a Three Year Period (After Lund and Ramsey, 1959) 120
7.1. Some Agencies, Such as the Air Force Manual 88-7 Use Equivalency Factors to Determine the Thickness Replacement Factor for Stabilized Layers. These Factors Are Normally Based on Strength 130
7.2. The Resilient Modulus of the Tama B Soil Is Significantly Influenced by Lime Stabilization Even After 10 Freeze-Thaw Cycles (After Thompson, 1985) 136
7.3. Lime Stabilization of the Vicksburg Clay Significantly Increases the Strength of the Soil and Hence the Stress Level at Which Permanent Deformation Will Accumulate (After TRB Report No. 5, 1987) 140
9.1. Example of Modern Heavy Duty Mixing Equipment Which Can Break Down and Pulverize to 457-mm (18-inches) in a Single Pass 163
9.2. Example of a Central Mixing Plant in Which Hydrated Lime Is Intimately Mixed with Soil to Reduce the PI 164
9.3. Lime Transport Trucks Used for the Delivery and Spreading of Dry Hydrated Lime 166
9.4. Pipe Spreader Bar Mounted at the Rear of a Tanker Used for Spreading Dry Hydrated Lime 167
9.5. Portable Type Spreader Attached to the Rear of a Truck Used for Spreading Dry Hydrated Lime 167
9.6. Mixing of Dry Hydrated Lime with Water in a Jet Mixer to Produce a Lime Slurry 169
9.7. Two Views of Portabatch Slaking Unit Used in the Production of Lime Slurry from Dry Quicklime. This Operation Was Used on the Massive Denver International Airport Project in 1991–1993 171
9.8. Disc-Harrows Used in Initial Mixing of Lime in a Two-Stage Mixing Process 175
9.9. Grader-Scarifiers Used in Initial Mixing of Lime in a Two-Stage Mixing Process 175

9.10. High-Speed Rotary Mixers Required for Final Mixing in a Two-Stage Mixing Process 176
9.11. Schematic Illustrating the Simple Basic Components Required for a Central Mixing Facility for Mixing Lime with Marginal Aggregate 177
9.12. The Level of Compaction Produced During Construction Has an Important Effect on In Place Strength of the Lime Stabilized Pavement Layer (After Thompson, 1969) 180
9.13. Sheeps-Foot Rollers Are Used to Compact the Lime-Treated Layer in One Lift. Rolling Is Continued Until the Roller "Walks Out" 181
9.14. After the Sheeps-Foot Roller "Walks Out" a Multi-Wheeled Pneumatic Roller Is Used to Complete Compaction to Meet Specifications. In Some Cases, Single Lift Compaction Can Be Accomplished With Heavy Pneumatic Rollers With Vibrating Impact Rollers 182
9.15. Moist Curing Is Provided by Keeping the Surface Damp by Sprinkling. This Is a Demanding Task in Dry Climates Or Under Dry Conditions 183
9.16. In Membrane Curing the Stabilized Soil Is Either Sealed With Cutback or Emulsified Asphalt at the Rate of Approximately 0.10 to 0.25 Gallons Per Square Yard Within One Day After Final Mixing. Subsequent Applications of Emulsion Are Often Required 184
12.1. LSPI Moves Through the Soil by Following the Paths of Least Resistance (After NLA Bulletin 331) 212
12.2. The Grid System for LSPI May Vary Depending on the Nature of the Project (After NLA Bulletin 331) 213
12.3. The Glazed Stabilized Compression Test Provides Information as to the Strength Gained by Glazed Coating of the Soil (After NLA Bulletin 331) 216
12.4. The Straight Seam Compression Test is Designed to Evaluate the Reinforcement Gained Through Vertical LSPI Seams (After NLA Bulletin 331) 216
12.5. The Angle Seam Compression Test is Designed to Evaluate the Reinforcement Gained Through LSPI Seams Which Occur at an Angle (After NLA Bulletin 331) 217

LIST OF TABLES

Table		Page
2.1.	Solubility of Lime at Different Temperatures Expressed in g/100g of Saturated Solution (After Boynton, 1979)	13
4.1.	Atterberg Limits for Natural and Lime-Treated Soils (After Little et al., 1987)	39
5.1.	Cured Strength Requirements For Soil-Lime Structural Layers (Modified from Thompson, 1970)	64
6.1.	CBR Values for Natural and Lime-Treated Soils (After Thompson, 1969)	80
6.2.	Compressive Strength Data for Natural and Lime-Treated Soils from Illinois, Texas and Colorado (Modified After Little et al. (1987))	85
6.3.	Change in Unconfined Compressive Strength During Curing for Twelve California Soils (After Doty and Alexander, 1978)	87
6.4.	Effective Roadbed Soil Resilient Modulus Values, M_R (psi), That May Be Used in the Design of Flexible Pavements for Low-Volume Roads. Suggested Values Depend on the U.S. Climatic Region and the Relative Quality of the Roadbed Soil (After AASHTO Pavement Design Guide, 1986)	100
6.5.	Back-Calculated Resilient Moduli of Aggregate Base Course Layers for Arizona Pavements	103
6.6.	Summary of Critical Strains in Phoenix, Arizona Pavements (After Little, 1990)	104
6.7.	In Place Resilient Modulus Calculation for Burleson Clay Subgrade at TTI Test Track	106
6.8.	Resilient Moduli of Lime Stabilized Subgrades (LSS's) Backcalculated from Falling Weight Deflectometer (FWD) Data (After Nowlin et al., 1992)	107
6.9.	AASHTO Structural Layer Coefficients Derived For Louisiana Base and Subbase Layers (After Van Til et al., 1972)	113
6.10.	Estimate of Bulk Stress (Θ) Induced in Aggregate Base Course as Influenced by Subgrade Support and HMAC Thickness (After AASHTO Pavement Design Guide, 1986)	114

6.11. Prediction of Aggregate Base Course Resilient Moduli Based on Moisture Content and Stress State (After AASHTO Pavement Design Guide, 1986) 114

6.12. Properties of Lime Stabilized Subgrade Below IH-20 Near Dallas, Texas, Illustrating Durability and Long-Term Strength Development 115

6.13. California Bearing Ratio Strength Summary for Selected Pavements. (After CERL Study, Aufmuth (1970)) 117

6.14. Field Test Data and Laboratory Test Data for Field Cores for South Dakota Highway No. 47 (After McDonald, 1969) 118

6.15. Comparison of South Dakota Pavements with Treated and Untreated Layers (Either Subgrade or Base Course Layers) (After McDonald, 1969) 118

6.16. Summary of Nebraska Pavements Evaluated in Deflection Study of Lime Treated Base and Subgrade (After Lund and Ramsey, 1959) 120

7.1. Strength Requirements for Lime Stabilized Bases, Subbases for Various State Agencies 132

7.2. In Situ Moduli Calculated from FWD Deflection Bases from SH 59 Near Houston, Texas 134

7.3. Approximation of Resilient Modulus from Unconfined Compression Strength Data. (After Terrel et al., 1979) 135

7.4. Comparison of Effects of Lime Stabilization on Two Clays: Tama B and Burleson, on the Resilient Modulus Response of an Overlying Aggregate Base Course (ABC) 137

7.5. Comparative Pavement Structures (Pavements A and B) 137

7.6. Summary of Important Mechanistic Parameters Used in Predicting Pavement Performance for Pavements A and B 138

7.7. Summary of Pavement Performance Prediction for Pavements A and B 138

7.8. Effect of Load Transfer Efficiency and Bonding on Subgrade Compressive Stress 140

8.1. Present Worth and Capital Recovery Factors (After Yoder and Witczak, 1976) 148

8.2.	Recommended Analysis Life for Comparing Pavement Alternatives (After Epps et al., 1987)	149
8.3.	Typical Life Cycles	151
8.4.	Average Life Cycles (After Corps of Engineers, 1987)	151
8.5.	Calculation Form for Present Worth Life Cycle Costing	154
8.6.	Example Problem Pavement Alternatives	155
8.7.	Modulus Values for Pavement Layers Used in Example Problem Pavement Alternatives	156
8.8.	Summary of Critical Design Parameters Used in Life Cycle Example	157
8.9.	Calculation of Percent Worth Life Cycle Costing	158

ACKNOWLEDGEMENTS

The author gratefully acknowledges the contributions of Dr. Jon Epps of the University of Nevada at Reno, Dr. Tom Petry of the University of Texas at Arlington, Dr. Marshall Thompson of the University of Illinois, Dr. Jim Eades of the University of Florida, and Dr. Tom Kennedy of the University of Texas at Austin. These gentlemen contributed considerably to the knowledge base upon which the handbook was developed. Special credit is given to Dr. Robin Graves of Chemical Lime Company for his critique and thorough technical review of the handbook as well as thoughtful contributions to the handbook.

The handbook was made possible through the support of the National Lime Association and especially the Texas Lime producers: APG Lime Company, Austin White Lime, Chemical Lime Company, Redland Stone Products and Texas Lime Company.

Mr. Ken Gutschick of the National Lime Association supplied the construction pictures.

PREFACE

Lime stabilization of clay soils has grown dramatically in usage since the Texas Highway Department conducted it's first trial job over 50 years ago. Today, lime stabilization has become widely used throughout the world with scientists in many countries contributing to the underlying scientific basis for the use of this stabilization method.

The list of scientists who have contributed to the technical support for this stabilization method is replete with names familiar to those in the geotechnical field. If any one man can be called the "father of lime stabilization", that man is Chester McDowell who did much of the pioneer work for the Texas Highway Department. He was active in the Transportation Research Board activities for many years, serving as Chairman of the Lime Stabilization Committee. Others who have made major contributions include Dr. Jim Eades of the University of Florida, who jointly with Dr. Ralph Grim of the University of Illinois developed the Eades-Grimm pH testing method now widely used and accepted as part of ASTM C-977 specification, covering lime for soil stabilization.

Other contributors include Dr. Marshall Thompson, of the University of Illinois, Dr. Jim Mitchell, of the University of California at Berkeley, E. B. McDonald of South Dakota, Ara Arman of Louisiana State University, Dr. Tom Petry of the University of Texas at Arlington, Dr. Tom Kennedy of the University of Texas at Austin and our author Dr. Dallas Little of Texas A&M University.

While contributions from these esteemed gentlemen and many others have advanced the use of lime stabilization of reactive soils, the need for a comprehensive handbook for use in training young engineers and for supplemental training for other engineers has long existed. Recognizing this need, the National Lime Association commissioned Dr. Dallas N. Little, Kelleher Professor of Transportation at Texas A&M University to develop such a handbook. Dr. Little has accepted the challenge by developing a comprehensive handbook. Not only has he brought together all of the pertinent data and facts about lime stabilization, he has also developed additional helpful data to further aid the engineer faced with stabilizing clay bearing soils.

It is our belief that this book conveys the subject better and in more depth than any other single publication at this time. We commend this book to you and thank Dr. Dallas Little for this major contribution to the geotechnical engineering field. We also wish to thank Dr. Robin Graves of the Chemical Lime Company for carefully reviewing the manuscript and for his valuable technical comments.

Thomas L. Potter
Executive Director
National Lime Association
Washington, D.C.

CHAPTER 1

INTRODUCTION TO THIS HANDBOOK

1.01 Purpose of the Handbook

This handbook provides a reference on the state of the art in lime stabilization of subgrade soils, subbases and base courses used primarily in roadway and airfield construction.

The handbook is not meant to be an exhaustive treatise on all aspects of lime stabilization. It should be supplemented with other authoritative sources such as National Lime Association publications, ASTM references, AASHTO references and specifications and procedures from various federal and state agencies as well as non-governmental agencies which use lime.

The handbook is designed to provide a comprehensive though not exhaustive reference on (1) mechanisms of reaction between lime and soil, (2) mixture design, (3) engineering properties derived as a result of lime stabilization of soils and aggregates, (4) pavement thickness design considerations, (5) construction and quality control considerations and (6) life cycle cost considerations.

1.02 Definition of Terms

Subgrade Stabilization

Subgrade stabilization includes stabilizing fine-grained soils in place (subgrade) or borrow materials which are employed as subbases, such as hydraulic clay fills or otherwise poor quality clay and silty materials obtained from cuts or borrow pits.

The stabilized in-place or borrow material is suitable for support of an overlying base course of varying thickness or a portland cement concrete slab or an asphalt surface layer. In most cases the enhanced engineering properties obtained as a result of lime stabilization can and should be included, or at least considered, in the pavement design and in the design of subsequent overlay courses.

The percentage of lime to be used for the treated subgrade must be determined by lab testing or empirical methods recognized in the literature. However, it is generally preferred that the optimum lime content be based on strength improvement and on ASTM C-977 which outlines the Eades-Grim method of verifying the proper amount of lime required for lime stabilization.

Base Stabilization

Base stabilization includes the upgrading of the strength and consistency properties of aggregates which may be considered unusable or marginal without stabilization. This includes aggregates with plastic fines, such as clay-gravel, "dirty" gravels and limestones, caliche and other marginal bases which generally contain an excess amount of fines passing the No. 40 mesh screen.

It is important that the applicability of lime stabilization to these marginal base materials is established through proper laboratory testing. This stabilization category applies to both new construction and reconstruction of distressed and worn-out roads and airfields.

Field mixing of the aggregate with a cutting and pulverizing machine is most commonly used for base stabilization, although central plant mixing may be more efficient, if economically justified.

Lime Modification

Lime modification of fine-grained soils or granular bases refers to improving the workability and constructability of soils or aggregates through the addition of small quantities of lime. The purpose of modification may be one or more of the following:

1. Aid compaction by drying out wet areas,
2. Bridge across underlying spongy subsoil,
3. Provide a working table for subsequent construction or
4. Condition the soil (making it workable) for further stabilization with fly ash, portland cement or liquid asphalt.

Because of the low percentage of lime used in lime-modification, the materials are less durable than lime-stabilized mixtures, but are nevertheless decidedly improved and are able to meet a more limited criteria. The main distinction between soil-lime modification and soil-lime stabilization is that generally no structural credit is accorded the lime-modified layer in highway design, such as a reduced pavement thickness.

Lime modified soils normally can be upgraded to lime stabilized soils through the addition of higher quantities of lime in accordance with mixture design criteria. The higher quantities of lime provide the critical amount of stabilizer necessary to drive the pozzolanic reactions required for strength development.

The Construction Procedure

The steps involved in stabilization or modification include: scarification and partial pulverization, lime spreading, wetting, mixing, compaction to maximum practical density and curing prior to placing subsequent layers or a wearing course.

1.03 Organization of the Handbook

The handbook is divided into 12 chapters.

Chapter 2 provides an introduction to lime, including definitions of pertinent terms, a description of the forms of lime available and most useful in roadbed stabilization, a discussion of important physical and chemical properties of quicklimes and hydrated limes, and a brief introduction to the production of lime and the lime industry.

Chapter 3 describes the role of the stabilized soil subgrade layer and/or the stabilized aggregate base layer in the pavement structure. The chapter discusses in very basic terms the importance of the improved properties (through stabilization) of the lime treated layer and the influence of the stabilized layer on improving the performance of each structural layer.

Chapter 4 defines the basic mechanisms of lime-soil stabilization and discusses these mechanisms. Not every reader is interested in an in-depth discussion of the reasons why stabilization works. However, it is important in a handbook of this type to offer this explanation to the interested reader. This basic understanding of reaction mechanisms may be important to the engineer facing specific challenges related to the stabilization of difficult soils.

Chapter 5 discusses the fundamental considerations involved in mixture design. Since a large number of mixture design approaches have been developed and are used by various agencies across the country, no one single approach is advocated. It is necessary for the reader to select the proper mixture design approach based on the agency for which the mixture design is being performed and for which the lime treated layer will be used.

Chapter 6 presents a broad-based discussion of the engineering properties of lime treated soils and aggregates. The chapter differentiates between uncured ("immediate") effects of stabilization and cured ("long-term") effects. Both consistency changes in the soil and strength changes are discussed. This chapter presents in situ performance and engineering property data as well as laboratory derived engineering property data. The durability and long-term performance of lime stabilized pavement layers are discussed and supported by laboratory and field data.

Chapter 7 addresses pavement structural thickness design considerations. No specific thickness design methodology is widely accepted for lime stabilized layers. Hence, this chapter offers suggestions and considerations for the thickness design of pavements containing lime stabilized layers.

Chapter 8 outlines an approach that can be used to consider pavement design strategies on a life cycle cost basis. An example is presented in this chapter where the performance of a pavement containing a lime stabilized subgrade is considered on a life cycle cost basis. The pavement is compared to an identical pavement but without the use of lime in the natural subgrade.

Chapter 9 discusses the construction procedure for lime stabilized bases, subbases and subgrades. The steps of the construction process are presented and discussed. The advantages and disadvantages of the use of dry hydrated lime versus quicklime and/or lime slurry are discussed. National Lime Association (NLA) Bulletin 326, "Lime Stabilization Construction Manual," should also be consulted as should Transportation Research Board Report No. 5—"State of the Art: Lime Stabilization," (1987).

Chapter 10 presents quality control steps that can and should be used in the field to insure a quality lime stabilization product. This chapter also lists the pertinent standards, specifications and other publications which should be referred to in the design of lime mixtures and in the construction of lime treated pavement layers.

Chapter 11 discusses the use of lime in recycling of pavements. Concentration is placed on in-place recycling of existing base and subbase layers and recycling of low volume roads.

Finally, *Chapter 12* presents a discussion of lime slurry injection and describes the injection process, injection materials, mechanisms of pressure injection stabilization, candidate soils for pressure injection and safety precautions. NLA Bulletin 331, "Lime Slurry Pressure Injection," should also be consulted.

1.04 Effective Use of the Handbook

If the objective of the reader is to overview lime stabilization, then Chapters 2, 3, sections 4.01 and 4.02 of Chapter 4 and section 6.01 of Chapter 6 are applicable.

For the reader whose purpose is to develop a more in-depth understanding of certain aspects of lime stabilization, the individual chapters provide this with a list of pertinent references to provide the reader with more detailed information.

1.05 Example Use of the Handbook

Lime is being considered as a candidate stabilizer for two soils: a highly plastic clay (CH) and a low plasticity clayey sand (SC). According to Section 4.06 of this handbook, both soils are candidates for lime stabilization based on their classification according to the Unified Classification System. Section 4.07 further identifies factors which may influence the degree and rate of pozzolanic strength gain when lime is used in these soils.

Based on the mineralogical, climatic and environmental factors discussed in Section 4.07, the engineer or designer decides that the potential for unfavorable reactions which may inhibit the stabilization process is absent or at least insignificant. Based on the recommendations presented in Section 4.08, the engineer or designer consults the U. S. Department of Agriculture's County Soil Report for information with regard to the important factors discussed in Section 4.07.

Based on the information presented in the County Soil Report, the engineer is aware that although the CH clay may be modified (constructability and volume change potential substantially improved and bearing capacity improved) with a relatively low percentage of lime, a higher lime content may be required to stabilize the soil.

Despite the relatively low clay content of the SC soil, the County Soil Report indicates that the clay mineral that makes up part of this soil is a mixed layered clay mineral containing the mineral smectite and is, therefore, considered reactive with lime. Thus the expectations are that significant strength improvement can be achieved through the stabilization of the SC soil with lime.

The Thompson mixture design procedure presented in Chapter 5 is used to determine the optimum lime percentage for both soils. This procedure is selected as it incorporates the Eades and Grim pH test to identify the amount of lime required to satisfy initial reactions and provide sufficient lime for long-term strength gain. This lime content is then further verified through strength testing according to the Thompson procedure.

Based on the properties of the natural soils and information from the mixture design, engineering properties of the mixture can be predicted based on the information presented in Chapter 6. The advantages provided by using these mixtures as a subbase are discussed in Chapter 3 and pavement design considerations are given in Chapter 7.

With the information provided in Chapter 6 (engineering properties) and the structural pavement design considerations discussed in Chapters 3 and 7, the engineer is able to estimate the potential benefits derived from incorporating the lime stabilized layer in the pavement structure. Finally, Chapter 8 provides the tools to perform a life cycle cost analysis by which to compare pavement alternatives.

Construction procedures and specifications and quality control of construction are presented in Chapters 9 and 10.

1.06 References

NLA Bulletin 326, (1991). *Lime Stabilization Construction Manual,* National Lime Association.

NLA Bulletin 331. *Lime Slurry Pressure Injection Bulletin,* National Lime Association.

Transportation Research Board, (1987). *State of the Art Report No. 5: Lime Stabilization.*

CHAPTER 2

INTRODUCTION TO LIME

2.01 Definition of Terms

Lime is one of the oldest and most versatile and vital chemicals known to man. Quicklime (calcium oxide) is produced by calcining high quality limestone at elevated temperature, volatilizing nearly half of the stone's weight in carbon dioxide. Hydrated lime (calcium hydroxide) is produced by reacting the quicklime with sufficient water to form a white powder. In this handbook, the term *lime* refers to calcium oxide (CaO) or calcium hydroxide ($Ca(OH)_2$) and *not* "agricultural lime" which is calcium carbonate ($CaCO_3$) (non-calcined). Lime may also include magnesium oxide or magnesium hydroxide. These forms of lime should be clearly designated as dolomitic lime. Dolomitic lime is less reactive with soil than is high calcium lime.

Several terms are used to describe different forms of lime in soil stabilization. It is important to understand the meanings of these terms from the outset so that no mistake is made by incorrectly interchanging one form of lime for another.

Lime is a general term that connotes a burned form of lime, usually quicklime, but may also refer to hydrated lime. It may be calcitic, magnesian or dolomitic. It does not apply to limestone or any carbonate form of lime. The production of lime begins with the calcination of limestone or dolomitic limestone. Stated chemically, this reversible reaction for both high calcium and dolomitic quicklime is diagrammed as follows:

Limestone + Heat → High Calcium Quicklime + Carbon Dioxide
$CaCO_3 + (\sim 1315°C) \rightarrow CaO + CO_2$
Dolomitic Limestone + Heat → Dolomitic Quicklime + Carbon Dioxide
$CaCO_3 \cdot MgCO_3 + (\sim 1315°C) \rightarrow CaO \cdot MgO + CO_2$

The hydration process which transforms quicklime into hydrated lime is as follows:

High Calcium Quicklime + Water → Calcium Hydroxide + Heat
$CaO + H_2O \rightarrow Ca(OH)_2 + Heat$
Dolomitic Quicklime + Water → Hydrated Lime + Heat
$CaO \cdot MgO + H_2O \rightarrow Ca(OH)_2 \cdot MgO$ or $Ca(OH)_2 \cdot Mg(OH)_2 + Heat$

There are three essential factors in the kinetics of limestone decomposition and hence in the production of lime:

1. The stone must be heated to the dissociation temperature of the carbonates,

2. This minimum temperature must be maintained for a sufficiently long duration and
3. The carbon dioxide gas that is evolved must be removed.

In addition to these factors other variables have a profound effect on the quality of the lime produced. These factors include: stone quality, stone size and gradation, rate of calcination, calcination temperature, calcination duration, chemical reactivity, stone density and porosity, surface area and spacing of the crystals and quality and type of fuel used in the calcination process. Thus, the production of high quality lime is not a simple task but must be monitored carefully in order to accommodate the changes in the nature of the limestone and other process variables (Boynton, 1979).

The lime, produced from the calcination of limestone, used and/or referred to in the soil and aggregate stabilization industry comes in many forms. These include:

Agricultural Lime—represents a relatively coarse, unrefined form of either hydrated lime or quicklime that is mainly used for neutralizing soil acidity and where lime of high purity and uniformity is not necessary.

Agricultural Limestone—is ground or pulverized limestone whose calcium and magnesium carbonate content is capable of neutralizing soil acidity but no calcination has taken place.

Available Lime—represents the total free lime (CaO) content in a quicklime or hydrated lime. This is the active constituent and is used as a means of evaluating the concentration of lime. This term is frequently used when comparing the reactivity of various chemical soil stabilizers with soil and aggregates. It is very important to understand the meaning of this term when assessing the relative merits of various stabilizers.

Carbide Lime—is a waste lime hydrate by-product of the generation of acetylene from calcium carbide and may occur as a wet sludge or dry powder of widely varying degrees of purity and particle size.

Dolomitic Lime—indicates the presence of from 35 to 46 percent magnesium carbonate in the limestone from which the material was formed.

High Calcium Lime—indicates the presence of 0 to 5 percent magnesium carbonate in the limestone from which the material was formed.

Hydrated Lime—is a dry powder obtained by hydrating quicklime with sufficient water to satisfy the chemical affinity, forming a hydroxide due to its chemically combined water. Hydrated lime may be high calcium lime, or dolomitic.

Lime Slurry—is a form of lime hydrate in a wet, suspension of solid lime, containing substantial amounts of free water.

Limestone—is a sedimentary rock consisting chiefly of calcium carbonate or of the carbonates of calcium and magnesium. Limestone may be of high calcium, magnesium or dolomitic form.

(1) Dolomitic Limestone—limestone containing from 35 to 46 percent magnesium carbonate.

(2) Magnesium Limestone—limestone containing from 5 to 35 percent magnesium carbonate.

(3) High Calcium Limestone—limestone containing from 0 to 5 percent magnesium carbonate.

Quicklime—is a lime oxide formed by calcining limestone so that the carbon dioxide is liberated. It may be high calcium, magnesian or dolomitic and of varying degrees of purity. Quicklime may come in many sizes from quite fine to very coarse.

Quicklime is available in a number of more or less standard sizes as follows:

Lump Lime—the product with a maximum size of 203-mm (8-inches) in diameter down to 51 to 76-mm (2 to 3-inches).

Crushed or Pebble Lime—the most common form, which ranges in size from about 51 to 6-mm (2 to 1/4-inches).

Granular Lime—the product obtained from Fluo-Solids kilns that have a particulate size range of 100 percent passing a #8 (U.S. Standard Sieve Size) sieve and 100 percent retained on a #80 sieve.

Ground Lime—the product resulting from grinding the larger sized material and/or screening off the fine size. A typical size is substantially all passing the #8 sieve and 40 to 60 percent passing a #100 sieve.

Pulverized Lime—the product resulting from a more intense grinding than is used to produce ground lime. A typical size is substantially all passing a #20 sieve and 85 to 95 percent passing a #100 sieve.

Pelletized Lime—the product made by compressing quicklime fines into about 25-mm (1-inch) sized pellets or briquettes.

Quicklime Slurry—is a mixture of quicklime and water in an exothermic reaction that produces hydrated lime in a slurry form. This slurry is comprised of a homogeneous blend of finely divided lime particles suspended in water.

Slaked Lime—is the hydrated form of lime, as a dry powder, putty or aqueous suspension. Only enough water is added to reach dry hydrated lime, versus an excess of water required to produce quicklime slurry.

Type S Hydrated Lime—(also called special hydrated lime) is an ASTM designation to distinguish a structural hydrate (designated type S) from a normal hydrated lime (designation type N). Type S hydrated lime possesses specified plasticity, water retention and gradation requirements. It may be dolomitic or high calcium and is more precisely milled than type N hydrates.

2.02 Forms of Lime Generally Available for Stabilization

Various forms of lime have been successfully used for soil stabilization. However, the most commonly used forms are hydrated high calcium lime, $Ca(OH)_2$; dihydrated dolomitic lime $(Ca(OH)_2 \cdot Mg(OH)_2)$; monohydrated dolomitic lime, $Ca(OH)_2 \cdot MgO$; calcitic quicklime, CaO; and dolomite quicklime, $CaO \cdot MgO$. Although the majority of lime stabilization in the United States utilizes hydrated lime, the use of quicklime accounts for more than 10 percent of the lime used in soil and aggregate stabilization.

Both high calcium lime and dolomitic lime have been successfully used for stabilization. However, certain basic physical and chemical differences do exist between high calcium lime and dolomitic lime that affect lime-soil reactivity. Among the most important of these are:

(1) $Ca(OH)_2$ is about 100 times more soluble than $Mg(OH)_2$ which means that high calcium lime provides more free calcium or available calcium for stabilization.
(2) MgO does not affect the solubility of $Ca(OH)_2$ but it may retard its rate of solution which, in turn, may affect the rate of reaction of the lime with the soil.
(3) MgO hydrates substantially more slowly than CaO; hence, the MgO in dolomitic quicklime may not fully hydrate prior to compaction of the stabilized soil. This could result in expansion after compaction as the MgO slowly hydrates with time.

2.03 Important Physical Properties of Quicklimes and Hydrated Limes

The physical properties of lime are discussed in detail by Boynton (1979). Generally quicklime is white of varying degrees of intensity, depending on its chemical purity. The purest quicklimes are the whitest. Impurities within the lime result in a grayish or yellowish appearance. Hydrated limes reflect a similar relationship between purity and whiteness. Quicklimes are invariably whiter than their derivative limestones and, in turn, hydrated limes are whiter than their derivative quicklimes.

As one would expect the crystal structures of quicklime and hydrated lime are different. X-ray diffraction (XRD) reveals that a pure calcitic oxide crystallizes in the cubic system. Hydrated lime is a hexagonal-shaped plate or prism with perfect basal cleavage, but the physical particle is of varying size since the microscopic crystallites agglomerate in varying degrees (Boynton, 1979).

Four of the most important physical properties of lime related to soil stabilization are: specific gravity, bulk density, heat of solution and solubility.

Specific Gravity and Bulk Density

The true specific gravity of pure calcium oxide is 3.34. This presupposes zero porosity. Values have been reported as high as 3.40 and as low as 3.0. Values for pure dolomitic lime, may range as high as 3.6. However, 3.34 appears to be a good average value (Boynton, 1979).

The apparent specific gravity is a more meaningful property because it represents the density of the actual material as it comes from the producer. Values of apparent specific gravity for quicklime vary from 1.6 to 2.8 with dolomitic quicklimes averaging from 3 to 4 percent higher. Average values of commercial oxides are 2.0 to 2.2. The range of specific gravities for different commercial hydrates are:

high calcium	2.3–2.4
highly hydrated dolomitic	2.4–2.6
normal hydrated dolomitic	2.7–2.9

The range of bulk density for quicklimes is from 768 to 1120 Kg/m^3 (48 to 70 pounds per cubic foot) with an estimated average of 880 to 960 Kg/m^3 per cubic foot (55–60 pounds per cubic foot) for pebble-sized quicklime. The bulk density for commercially available hydrates ranges from 400 to 640 Kg/m^3 (25 to 40 pounds per cubic foot) with an average value of approximately 564 Kg/m^3 (35 pounds per cubic foot) (Boynton, 1979).

The molecular weight of CaO is 56.08 (Boynton, 1979), and the molecular weight of Ca(OH)$_2$ is 74.10. Based on this ratio of molecular weights, it is apparent that in order to provide equal levels of CaO, it will require more Ca(OH)$_2$ than CaO. This is because of the extra H$_2$O attached to each molecule of calcium hydroxide or hydrated lime. Based on the ratio of molecular weights, 32 percent more calcium hydroxide or hydrated lime is required than calcium oxide or quicklime in order to provide equal amounts of calcium oxide and hence available lime, which is critical for lime-soil reactions. Of course the substitution ratio between hydrated and quicklime based on available lime varies somewhat depending on purity of the quicklime and hydrated lime. Most specifying agencies prefer hydrated lime or quicklime slurry (which is hydrated lime slurry) over dry quicklime. If dry quicklime is permitted by specification, most experienced agencies require the same amount of quicklime as hydrated lime because of the undependable nature of quicklime slaking in the soil.

Particle Size

The typical particle sizes of quicklime are discussed under **Definition of Terms**. Hydrated lime is air classified to produce the fineness necessary to meet the requirements of the user. The normal grades of hydrate used for chemical purposes will have 75 to 95 percent passing the #200 sieve; while for special uses the hydrate may be classified as fine as 99.5 percent passing a #325 sieve. Due to air classification, generally the commercial hydrate produced is purer than the quicklime from which it is derived since much of the impurities are rejected in the classifier.

Heat of Formation

The heat of formation is synonymous with the heats of hydration and reaction. For commercial hydrates, the heat of hydration for $Ca(OH)_2$ is approximately 6.396×10^7 $J/(kg \cdot k)$ (27,500 Btu/lb mole) or 1.14×10^6 $J/(kg \cdot k)$ (488 Btu/lb mole) of quicklime. For $Mg(OH)_2$ the values are somewhat lower, 3.35×10^7 to 4.2×10^7 $J/(kg \cdot k)$ (14,400 to 18,000 Btu/lb mole). This substantial heat of hydration is important in the production of hydrated lime or quicklime slurries (Boynton, 1979).

Solubility of Hydrated Lime

The solubility of $Ca(OH)_2$ is 1.330 g CaO/l of saturated solution at 10°C in distilled water. At 0°C solubility increases to 1.4 g CaO/l. $Ca(OH)_2$ is approximately 100 times more soluble than calcium carbonate. The solubility of lime expressed as CaO or $Ca(OH)_2$ at different temperatures in g/100g of saturated solution is presented in Table 2.1.

The solubility of $Ca(OH)_2$ is affected by some salt and inorganic chemical solutions in varying degrees, depending on concentrations. Most salts increase the solubility of hydrated lime by about 10 to 15 percent. Generally, increases in temperature still depress solubility of the hydrate. Specific information concerning the effects of different salts, organic solutions and other impurities on the solubility of $Ca(OH)_2$ and $Mg(OH)_2$ is discussed by Boynton (1979).

Table 2.1. Solubility of Lime at Different Temperatures Expressed in g/100g of Saturated Solution (After Boynton, 1979).

Temperature, °C	Solubility of CaO, g/100g	Solubility of $Ca(OH)_2$, g/100g
0	0.140	0.185
10	0.133	0.176
20	0.125	0.165
30	0.116	0.153
40	0.106	0.140
50	0.097	0.128
60	0.088	0.116
70	0.079	0.104
80	0.070	0.092
90	0.061	0.081
100	0.054	0.071

2.04 Important Chemical Properties of Quick and Hydrated Limes

Quicklime and hydrated lime are reasonably stable compounds. However, quicklime is vulnerable to water; even the moisture in the air produces a destabilizing effect by air slaking. Hydrated lime is more stable since water does not cause a change in its composition. The primary factor influencing the stability of hydrated lime is carbon dioxide which reacts with either quicklime or hydrated lime to form calcium carbonate (Boynton, 1979), generally at a slow rate.

The reactivity of quicklime with water is of great practical importance as this reactivity is the basis for the production of hydrated lime from quicklime. The production of hydrated lime from quicklime through a slaking process at a construction site produces a very reactive product for soil stabilization.

Obviously, a relationship exists between hydrated lime and the quicklime from which it was derived. The chemical composition of the hydrated lime reflects this relationship. A high calcium quicklime will produce a high calcium hydrated lime containing 72 to 74 percent calcium oxide and 23 to 24 percent water in combination with the calcium oxide. A dolomitic quicklime will produce a dolomitic hydrate. Under normal hydrating conditions the calcium oxide fraction of the dolomitic quicklime completely hydrates, but generally only a small portion of the magnesium oxide hydrates (about 5 to 20 percent). The composition of a normal dolomitic hydrate will be 46 to 48 percent calcium oxide, 33 to 34 percent magnesium oxide and 15 to 17 percent water in chemical combination with calcium oxide (Boynton, 1979).

pH of Lime-water Solutions

The pH of solutions at 25°C (77°F) rise sharply with the addition of very low concentrations of $Ca(OH)_2$. A concentration of only approximately 0.064 g/l of hydrated lime will increase the pH of distilled water from 7 (neutrality) to above 11. From this point the pH rise with increased hydrated lime concentration is gradual. The pH of the solution peaks at approximately 12.454 at 25°C (77°F). Temperature is an important factor since rises in temperature reduce solubility of $Ca(OH)_2$ and, therefore, decrease the pH slightly. Since $Mg(OH)_2$ is substantially less soluble than $Ca(OH)_2$, more dolomitic hydrate than $Ca(OH)_2$ is required to reach the peak pH of the solution.

The high pH of a lime-water solution is of great practical importance in soil and aggregate stabilization. This is because a high pH or basic environment increases the ability of the lime to react with soil minerals, and produce a cementitious product which can stabilize particles by "gluing" them together (Boynton, 1979).

Rate of Solution

Lime in solution produces a high pH environment which can react with soil minerals. The rate and efficiency of this reaction between lime and water depends on many factors. But, perhaps the most important factors in soil stabilization are the particle size of the lime and the nature of the solute.

The rate of solution of hydrated lime in water is heavily influenced by particle size. As the average diameter of the particle's size diminishes, its dissolution increases, since its surface area is augmented.

Statistically, it can be stated that the rate of solution of hydrated lime increases with higher specific surface areas of hydroxide particles.

The effect of organic solutes, such as sugar, acids and salts, have a profound effect on lime solubility and the rate of solution. An extensive treatise of these effects is presented by Boynton (1979).

Reaction of Lime With Carbon Dioxide

The reaction between lime, either in the quicklime form or in the hydrated form, and carbon dioxide results in the reformation of calcium carbonate ($CaCO_3$). This process should be minimized during construction as it robs the system of lime in the

CaO or $Ca(OH)_2$ form, which is reactive with the soil minerals. However, complete carbonation usually does not occur, even at elevated temperatures. This is because the adsorption of CO_2 on the lime is a surface phenomenon, and a shell of calcium carbonate is gradually formed around the CaO or $Ca(OH)_2$ particle (Boynton, 1979).

It is, however, important to note that water acts as a catalyst for carbonation, and the process of carbonation is more complete on small-sized, high surface area particles of lime than on larger or coarser-sized particles.

Reaction With Silica and Alumina

Lime reacts with many compounds and elements including sulfur compounds, acid gases, halogens, magnesia compounds, iron, phosphorous, silica and alumina and metals (Boynton, 1979).

The most important reaction among these in terms of soil and aggregate stabilization is the reaction between lime and silica and alumina compounds. The reactions between lime and available silica and alumina are quite complex, and there is not complete agreement on all aspects of the systems that form as a result of these reactions. However, it has been well documented that CaO reacts with silica (SiO_2) and alumina (Al_2O_3) and water at elevated temperatures (93°C to 260°C, 200°F to 500°F) to form hydrated calcium silicate and calcium-aluminate compounds (Boynton, 1979).

Eades and Grim (1965) demonstrated through X-ray diffraction and differential thermal analysis that complex calcium silicates are formed by the reaction of lime and the available silica contained in clay minerals (kaolinite, montmorillonite, illite and chlorite) when compacted under optimum moisture conditions and exposed to either normal climatic and temperature conditions or laboratory curing.

Eades (1962) reported that the formation of the calcium silicate hydrate compounds were the result of the lime attacking the edges of the clay minerals in the high pH environment of the lime-water solution. The lime actually eroded these microparticles with formation of noncrystalline gelatinous calcium silicates that behave like a cement in binding these particles together.

Environmental Effects

Lime is not toxic to workers in construction, manufacturing or lime consuming plants, nor are air-borne dust particles harmful to the public. After steel making, lime's greatest use is for environmental cleanup of water, wastewater, air and solid wastes. However, because of the high alkalinity of lime products, safety procedures and equipment are encouraged as outlined in National Lime Association Bulletin 326.

2.05 Production of Lime

Lime must meet exacting chemical and physical specifications and, thus, high purity limestone is required. The processing of stone to make lime is carefully controlled to insure a quality product. Four basic steps are followed: quarrying and mining, stone preparation, calcining and hydrating (Boynton, 1979).

The quarrying and mining operations include overburden removal, drilling, blasting and transportating to the plant. Stone preparation refers to the primary and secondary crushing and screening operations necessary to produce desired gradations of stone for kiln feed purposes. Prior to feeding the kilns the stone often is washed to minimize contamination.

Almost 90 percent of lime is made in giant rotary kilns, Figure 2.1, ranging up to nearly 154-m (500-ft.) long and 5.2-m (17-ft.) in diameter and producing over 1,089 metric tons per day. Vertical (shaft) and other special kilns comprise the balance. Lime calcining is energy intensive, requiring up to 0.27 metric tons of coal to produce one

FIGURE 2.1. ROTARY KILNS ARE USED IN THE PRODUCTION OF LIME.

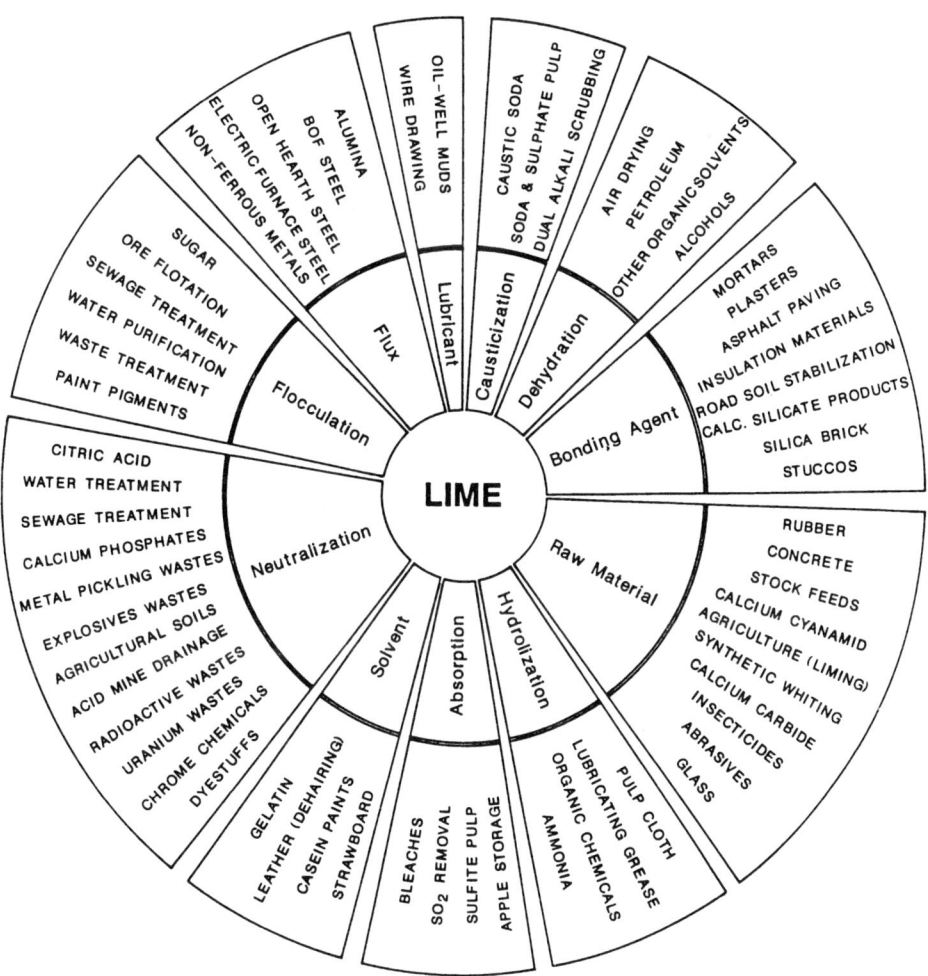

FIGURE 2.2. LIME HAS MANY USES IN INDUSTRY (AFTER NLA BULLETIN 214, 1992).

ton of lime. After calcining, the quicklime is cooled, crushed if necessary and stored for shipment.

For many manufacturers, the next logical phase of lime production following calcining to produce quicklime is to produce hydrated lime by reacting the quicklime with water in continuous hydrators. The dry hydrate, consisting of micrometer-sized particles, is then classified by air separators, which reject coarse particles. Hydrate is then stored and shipped in bulk or in bags.

2.06 Lime Industry

Figure 2.2 depicts the leading use categories for lime, based on annual U. S. production of nearly 19 million metric tons (National Lime Association Bulletin 214, 1992).

The dominant role of lime is in steel production. Here lime serves as a flux for removing impurities in refining steel. In basic oxygen furnace (BOF) steel, an average of 66 kg (140 lb.) of quicklime is consumed per ingot ton.

In non-ferrous metallurgy, lime is used to beneficiate copper ore, make alumina and magnesia for use in aluminum and magnesium manufacture, extract uranium and recover gold, silver, etc. The chemical industry requires lime to make such chemicals as sodium alkalies, calcium carbide, calcium hypochlorite, citric acid, petrochemicals, etc. The paper industry uses lime as a causticizing agent and for bleaching. In construction, lime's traditional use in mortar and plaster still flourishes. However, lime's largest construction use is in the stabilization of roads, airfields, building foundations, earth dams, etc., where it upgrades low quality clayey soils into satisfactory base and subbase materials. An allied use of lime in construction is as an addition to hot mix asphalt concrete. Lime improve durability of hot mix by improving resistance to moisture damage and reducing oxidative aging in the asphalts cement. Other uses include refractories, sugar refining, agricultural liming, glass making, leather tanning, plastics and pigments, and many other uses.

The largest environmental use of lime is potable water softening and clarification. In wastewater treatment of sewage effluents, lime removes phosphorus and nitrogen; abets clarification; produces a high pH environment unsuitable for the growth of bacteria and virus; neutralizes acid mine and industrial wastewater discharges; and absorbs and neutralizes sulfur oxides from industrial stack gases, beneficiating air quality and stabilizes sludges from sewage and desulfurization plants for safe land disposal or for utilization in farming or construction.

2.07 References

Boynton, R. S., (1979). *Chemistry and Technology of Lime and Limestone,* John Wiley and Sons, Inc., New York.

Chemical Lime Facts (1992). National Lime Association Bulletin No. 214.

Eades, J. L., (1962). "Reactions of $Ca(OH)_2$ with Clay Minerals in Soil Stabilization," Ph.D. Dissertation, University of Illinois.

Eades, J. L., and Grim, R. E., (1962). "Reactions of Hydrated Lime with Pure Clay Minerals in Soil Stabilization," *Highway Research Board Bulletin,* No. 262.

CHAPTER 3

ROLE OF STABILIZED LAYERS IN THE PAVEMENT STRUCTURE

3.01 Role of the Surface Course

A pavement surface course is generally one of two forms: portland cement concrete (PCC) or asphalt concrete. When the surface is composed of PCC, the pavement is often referred to as a "rigid" pavement. When the surface is comprised of asphalt, either in the form of a seal coat or in the form of hot mix asphalt concrete (HMAC), the pavement is typically referred to as a "flexible" pavement. In either case the purpose of the surface is to provide a smooth, safe riding surface.

Portland cement concrete surfaces require uniform support to function properly and economically. These pavements are usually jointed and reinforced to accommodate shrinkage and temperature induced stresses. Portland cement concrete pavements will crack to some degree to release stresses developed by temperature fluctuations or drying induced shrinkage. Adequate long-term load transfer across the joints designed to accommodate shrinkage or across the cracks developed in response to shrinkage and/or temperature fluctuations requires uniform and stable support of the subbase directly below the PCC slab. Lime or other stabilizers can be used to help provide this type of support.

Asphalt surfaces can either be in the form of HMAC or a surface treatment or seal coat. These surface treatments essentially do not provide structural support but provide a smooth riding surface with adequate skid resistance to meet safety requirements. In addition, asphalt surfaces provide protection of the underlying layers from surface moisture.

Seal coats can be single or multiple chip seals, slurry seals or other microsurfacing techniques.

3.02 Role of Base and Subbase Courses

As previously stated, the role of the subbase under PCC pavements is to provide uniform and stable support for the PCC surface, Figure 3.1. In order to provide uniform support, the subbase must be either a free-draining material or resistant to erosion and hence resistant to pumping through joints and/or cracks. The action of erosion and subsequent pumping may lead to the loss of material directly below the joint or crack and lead to loss of support which, in turn, leads to further cracking and joint deterioration.

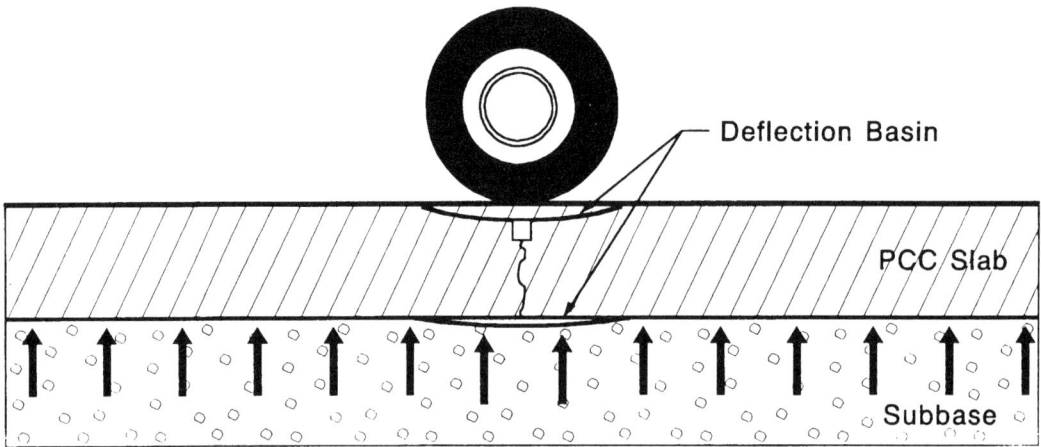

FIGURE 3.1. SUBBASE BELOW PCC PAVEMENT PROVIDES UNIFORM SUPPORT AND AIDS IN LOAD TRANSFER.

The subbase has *not* traditionally been considered to be a major structural layer in PCC pavements. In fact, it is customary to label the layers of PCC pavements as surface, subbase and subgrade. The absence of the "base" layer indicates that the dominant structural component is the PCC slab. However, the need for the subbase under the PCC slab to provide uniformity of support and resistance to permanent deformation has never been challenged. Recent work by Zollinger (1991) illustrates this point and is discussed in Chapter 6.

The base and subbase course are major structural components of a flexible pavement. Figure 3.2 illustrates the critical stresses in a typical flexible pavement with an HMAC surface. In this figure the HMAC is supported by an unbound aggregate base course (ABC) which is supported by a lime stabilized subgrade (LSS).

The flexural tensile stresses and strains induced in the asphalt concrete layer by traffic loads are related to fatigue cracking. If these stresses and strains are too large, the result is a pavement with asphalt cracking in the wheel path. The shearing stresses in the HMAC are related to distortion, shoving or rutting within the surface layer. These stresses are induced by the vertical contact stresses of the wheel load as well as the rolling and braking surface shearing stresses. The vertical compressive stresses and strains within the granular bases, subbases and subgrades are related to rutting and pavement roughness developed in the base and subbase layers.

3.03 Role of the Base and Subbase in Minimizing Pavement Distress in Flexible Pavements

The magnitude of the stresses and strains developed within the various pavement layers is dependent upon the magnitude of the contact stresses of the wheel load, the thicknesses of the respective layers and the relative stiffness or moduli of the various layers.

Since the magnitude of the stresses developed within the asphalt layer is heavily influenced by the modulus ratio (or stiffness ratio) between the surface and the base course, it is not surprising that strengthening the base or subbase is a method by which to minimize distress within the HMAC. An example of this is that by increasing the stiffness or modulus of a granular base by adding a stabilizer, such as lime, the ratio of moduli between the HMAC and base layer is reduced. In other words, the increased stiffness of the base provides better support for the surface. The net result is that both the *tensile flexural stresses* and the *shearing stresses* developed within the HMAC are reduced. The reduction in flexural and shearing stresses reduces the potential to fatigue crack or deform (Figure 3.3).

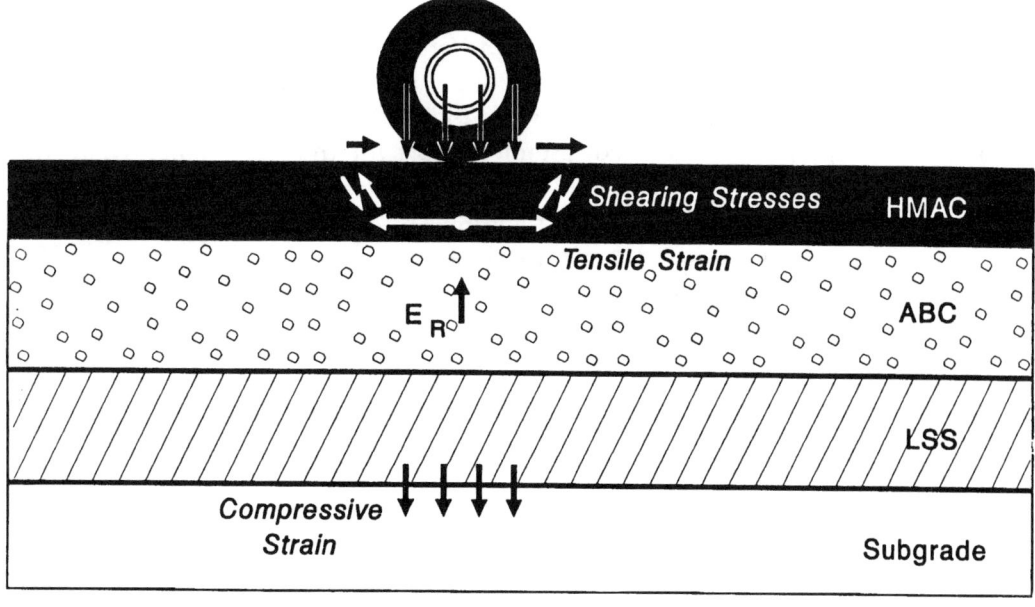

FIGURE 3.2. CRITICAL PARAMETERS IN A FLEXIBLE PAVEMENT INCLUDE FLEXURAL TENSILE STRAIN IN THE HMAC, SHEARING STRESSES WITHIN THE HMAC AND VERTICAL COMPRESSIVE STRESSES AND STRAINS AT THE TOP OF THE SUBGRADE.

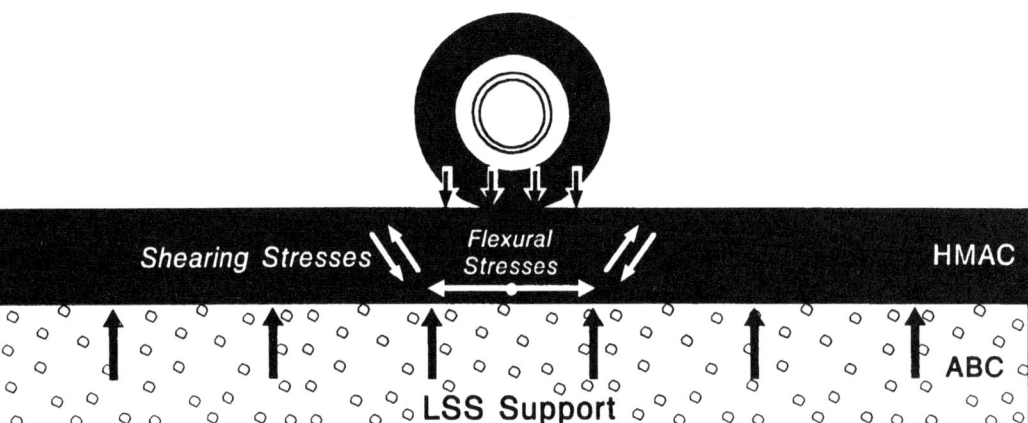

FIGURE 3.3. TENSILE STRAINS AND STRESSES IN THE HMAC ARE REDUCED AS A RESULT OF IMPROVED SUPPORT OF THE HMAC AND AGGREGATE BASE COURSE BY THE STABILIZED SUBGRADE LAYER.

The stiffness or modulus of granular bases is stress dependent. This concept is discussed in Chapter 6. In general this means that as the stress or confinement within the granular base is increased, the response modulus or stiffness increases. Development of a high confining stress within the granular base can be achieved by improved support of the subbase or subgrade. Since lime stabilization of a native soil improves the consistency of the soil over a wide range of moisture contents, it reduces plasticity, reduces swell potential and volume change potential and increases strength of the subgrade or subbase. This means that a lime stabilized subgrade or subbase can provide improved support and more consistent support for the base compared to the unstabilized, natural soil. The net result is a better structural response of the granular base and a reduction of the cracking and shearing distress within the HMAC.

Strengthening and stiffening the base and subbase layers through lime stabilization provides improved load-spreading capability of the pavement structure and hence protects the natural subgrade from being overstressed by traffic loading. The net result here is that the potential to develop pavement roughness or deep layer rutting is reduced, (Figure 3.4).

3.04 Important Characteristics of Lime-Stabilized Bases, Subbases and Subgrades to Meet Structural Pavement Demands

In order to perform acceptably, stabilized bases must provide adequate support for the HMAC surface and the load-distributing power expected to protect the underlying subgrade layer from being overstressed. The engineering property associated with these

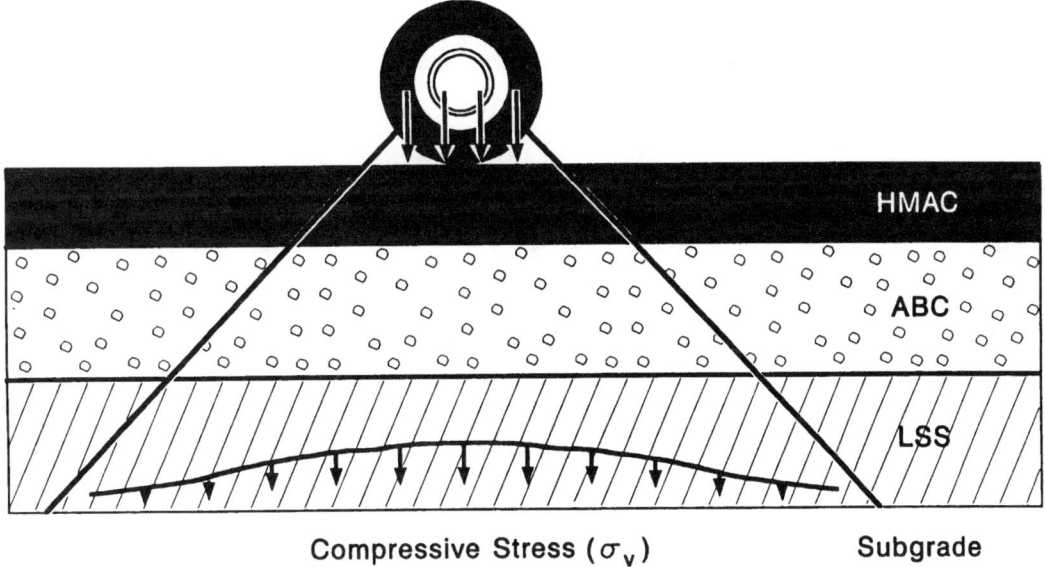

FIGURE 3.4. THE DISTRIBUTION OF VERTICAL COMPRESSIVE STRESSES AND STRAINS ARE CONTROLLED BY THE STIFFNESSES OR MODULI OF THE BASE AND SUBBASE LAYERS.

performance characteristics is the resilient modulus of the stabilized base. The resilient modulus is determined in accordance with AASHTO test method T-274. The resilient modulus of granular materials is the ratio of repeated stress applied to the specimen or pavement section to the measured total resilient or recoverable strain caused by the applied dynamic or cyclic load:

$$E_r = \sigma_{repeated} / \epsilon_{total\ recoverable}$$

where E_r is the value of the resilient modulus, $\sigma_{repeated}$ is the repeated stress which simulates the stress level produced by a moving wheel and $\epsilon_{total\ recoverable}$ is the measured total resilient strain induced as a result of the repeated stress.

Figure 3.5a illustrates the resilient modulus measured in the lab on a cylindrical sample loaded axially. Axial loading is intended to mimic the moving wheel load in the field, Figure 3.5b. The laboratory test attempts to duplicate field stress states by applying representative axial stresses and representative confining stresses during the test.

In addition to providing an acceptable level of stiffness or resilient modulus to properly distribute loads and to properly support the HMAC surface layer, the stabilized base must possess adequate tensile strength and shearing strength to resist fracture and distortion or deformation. Adequately high tensile strengths provide the ability of

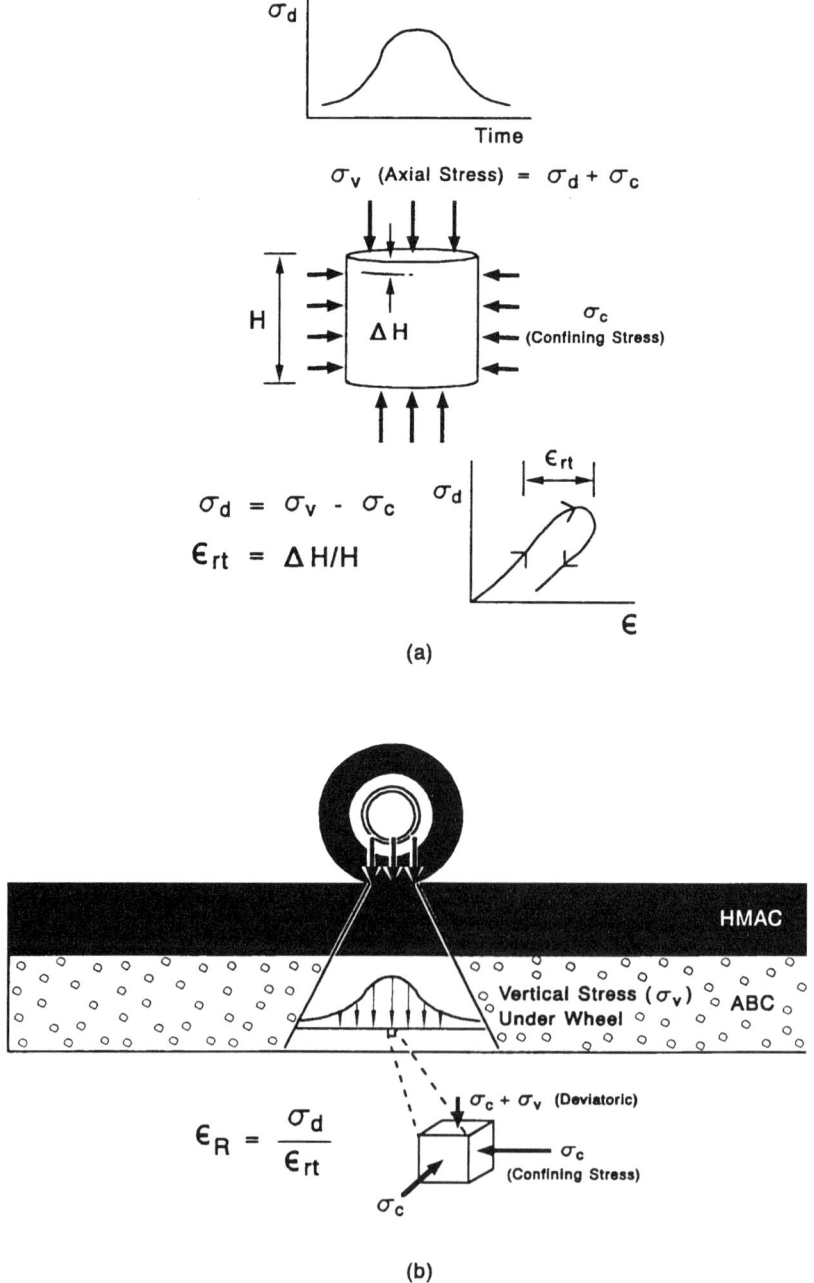

FIGURE 3.5. THE RESILIENT MODULUS TEST IS A PRIMARY METHOD OF CHARACTERIZING PAVEMENT MATERIALS FOR USE IN PAVEMENT LAYERS. THE LABORATORY TEST (A) ATTEMPTS TO SIMULATE FIELD LOADING CONDITIONS (B).

the stabilized base or subbase to resist traffic-induced load or environmentally induced tensile stresses which can result in cracking. Knowledge of the compressive strength of the lime-soil mixture is a necessary indicator of the ability of the base to resist shear failure. The engineering test properties associated with this required level of strength are the unconfined compressive strength, the indirect tensile strength and the flexural strength.

The unconfined compressive strength is a rather simple mixture property to measure and is by far the most widely used property by which to evaluate mixture strength. Fortunately, the tensile and flexural strength can be approximated from the unconfined compressive strength with acceptable accuracy for most pavement design and analysis considerations. The tensile strength is approximately 13 percent of the unconfined compressive strength while the flexural tensile strength is approximately 25 percent of the unconfined compressive strength.

In pavement design protocols, the potential of a pavement to exhibit fatigue cracking in the HMAC is typically determined by predicting the magnitude of the critical tensile strain within the HMAC under the load of the design vehicle and relating this to fatigue cracking by means of a transfer function. This transfer function is an empirical model which relates observed fatigue cracking damage in field pavement sections to a calculated mechanistic parameter, in this case tensile strain in the HMAC. Similarly, transfer functions are used to predict rutting from vertical compressive stresses in the granular layers and from shearing stresses in the HMAC.

The stiffness or resilient modulus for stabilized bases is sometimes great enough relative to that of the supporting subgrade so that the tensile stresses induced within the stabilized base under traffic loading are great enough to potentially cause fracture or fracture fatigue in the stabilized layer. A design approach to guard against fatigue cracking in stabilized bases is to insure that the calculated maximum tensile stress in the stabilized base is not high enough to induce fatigue. This is normally done through keeping the stress ratio (ratio of induced flexural stress to flexural strength) to less than 50 percent. Since the tensile flexural strength of a stabilized material is approximately 25 percent of the unconfined compressive strength, the maximum allowable load-induced tensile stress in the stabilized layer should not exceed 12.5 percent of the unconfined compressive strength of the stabilized layer, Figure 3.6.

3.05 Use of Lime in Upgrading Subgrade Soils and Aggregate Bases

By adding the appropriate quantity of lime to subgrade soils which are suitable for lime stabilization, the engineering properties of these soils can be upgraded. The improved engineering properties include:

1. *Reduced shrink—swell* (volume change) potential,

2. *Increased compressive, tensile and flexural strength* and
3. *Increased stiffness* or resilient modulus.

These improvements result in improved performance of these lime treated layers in the pavement system. The stronger, stiffer and more stable (volumetrically) lime treated subgrade, subbase or base layers provide better protection for weak and deformation susceptible natural subgrades and better support of unbound aggregate bases and asphalt surfaces, thus enhancing their performance.

Lime can be used to improve the strength and performance of good quality to high quality aggregate bases. It can also be used to upgrade marginal aggregates to meet specification requirements as acceptable subbases and bases. Lime is widely used in marginal aggregates to ameliorate the plastic clay content of marginal bases and to remove clay film from aggregates used as concrete aggregate, asphalt aggregate or in the ballast or sand industries.

The improvements derived from using lime to upgrade marginal aggregates are:

1. Plasticity of treated aggregate is reduced to a non-plastic or acceptable state,
2. Clay content is stabilized and becomes an important and marketable part of the aggregate,

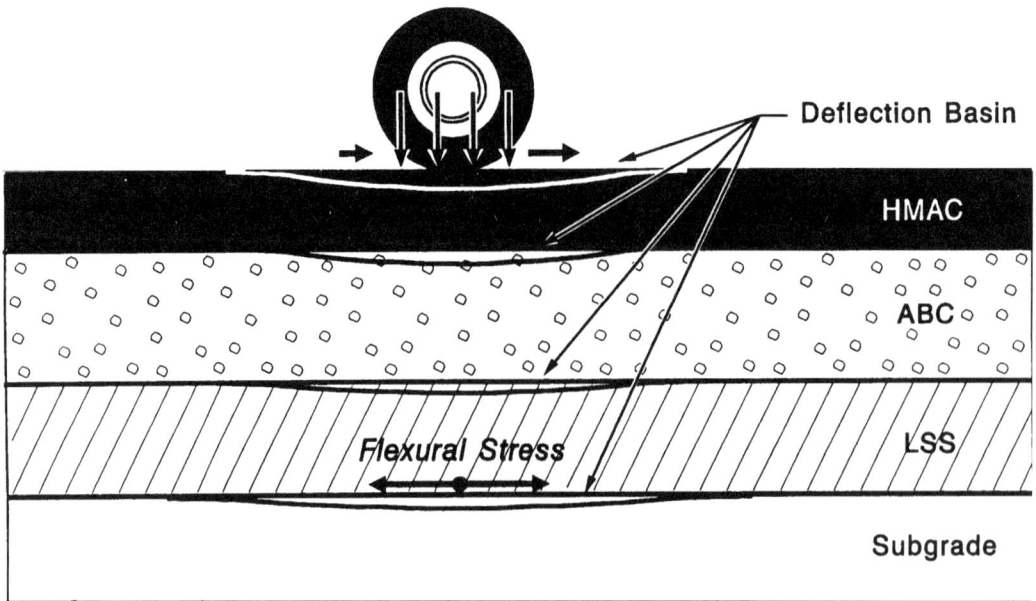

FIGURE 3.6. STIFF (HIGH MODULUS) STABILIZED SUBBASES OR BASES CAN DEVELOP HIGH FLEXURAL STRESSES WHICH SHOULD BE CONSIDERED IN LAYER THICKNESS DESIGN.

3. Strength gain is developed through pozzolanic reactivity and
4. Production life of certain sand and gravel pits and crushed stone quarries can be extended for *years* by converting highly plastic clay content segments into market quality raw materials.

The amount of lime needed to modify the deposit varies with the percentage of clay present in the material, and the plasticity of that clay. The first step in the treatment process is to obtain a good, representative sample of the material in the deposit, preferably collected by a soils engineer. The second step is to establish the plasticity of the clay in the representative sample. The next requirement is to establish the quantity of lime necessary to reduce the plastic index (PI) to an acceptable level, increase the sand equivalent to an acceptable level and to achieve the strength required, if applicable.

The quantity of lime required in marginal aggregate modification is usually in the range of 1.0 to 1.5 percent for aggregates with PI's below 20 and from 1.0 to 3.0 percent for aggregates with PI's above 20.

3.06 References

Zollinger, D., Kadiyala, S. and Smith, R. (1989). "Review of Alternate Pavement Design for Area A of Taxiway B and C Construction," Prepared for Dallas/Fort Worth International Airport, Report No. 4,100-1, Texas Transportation Institute.

CHAPTER 4

MECHANISMS OF LIME-SOIL STABILIZATION

4.01 General

Lime is an effective stabilizer for a wide range of soils. Actually, two phases of stabilization occur in a lime-soil system. The first involves the practically immediate reactions of cation exchange and flocculation-agglomeration. These reactions occur to some extent with all fine-grained soils. Due to textural changes caused by these reactions within the soil, the strength and moisture stability of these soils is improved. These improvements are reflected in improved workability, immediate strength improvement and reduced swell-susceptibility.

Pozzolanically induced long-term strength gain is more capricious as its success depends upon a reaction between the lime and the clay as well as the mineralogy of the clay. Many clays are reactive and upon lime stabilization their strength may easily triple or quadruple. In some instances strengths have improved by an order of 10 or more.

Lime, the versatile soil stabilizer, should be considered with all soils when the PI exceeds 10 and the percent of soil smaller than the number 200 sieve exceeds 25 percent.

4.02 Need for Stabilization

Soils often require stabilization to add mechanical stability, to improve durability or to alter their volume change potential. The most widely recognized form of stabilization is compaction, which improves the mechanical stability of virtually any soil. However, compaction alone is often not enough. This is especially true with fine-grained, cohesive soils.

Plastic clays pose a unique problem to the engineer. Their consistency varies over a very wide range. This consistency is directly related to the availability of water. The ability of clays to take-on and lose water is a function of their morphology and mineralogical nature.

Lime (either slaked quicklime or hydrated lime, in both high calcium and dolomitic types) is an effective stabilizer of clays. The lime actually alters the ability of the clay to hold water at its surface and can react with the clay to produce a cement which may add substantially to the strength of the lime-stabilized clay.

This chapter discusses the unique phenomena which occur when lime and clay are mixed. These phenomena result because of the unique mineralogy of clays and the chemical properties of the calcium and/or magnesium compounds present in the lime.

Soils are normally divided into groups according to particle size. In fact, the Unified Soil Classification System divides soils into categories of gravels, sands, silts and clays, based on size. A look at the mineralogy of soils, however, shows that differentiation according to size is only a beginning in understanding soil behavior.

4.03 Nature of Soil (Soil Mineralogy)

A course in soil mineralogy provides an excellent background for an engineer seeking to understand the phenomena involved in soil stabilization. The subject of soil mineralogy is complex and intricate. However, a simple comparison of the mineralogical structure of a few commonly encountered soils will illustrate the critical role of mineralogy and crystal structure.

The Earth's crust to a depth of about 10 miles is composed mostly of oxygen (47.3%), silicon (27.7%) and aluminum (7.8%). These are followed by smaller amounts of the metals iron (4.5%), calcium (3.5%), sodium (2.5%), potassium (2.5%) and magnesium (2.2%) (Mitchell, 1976). The manner in which these elements combine to form the compounds which form the soils and rocks in the crust is affected by the thermodynamics of the genesis process. Organized atomic arrangements don't just happen but occur in a fashion which will preserve electrical neutrality, satisfy bonding directionality, minimize strong ion repulsions and, in short, provide the most stable arrangement of atoms possible.

Two of the most common building blocks are shown in Figure 4.1. These are the *silica tetrahedron* and the *aluminum octahedron*. Within each building block the multivalent cation is coordinated with oxygen. Thus the elements comprising approximately 83 percent of the earth's crust are accounted for by these two basic building blocks.

The silica tetrahedron represents the basic coordination polyhedron which forms the widely abundant minerals quartz and feldspar. The silica tetrahedron is not electrically neutral. Tetrahedra link together in arrangements which minimize strong repulsions between the silicon ions. The high positive charge of the silicon ion develops a variety of possible packing arrangements in response to the repulsions generated between adjacent cations. Figure 4.2 shows one such arrangement of these tetrahedra—three-dimensional space lattice silicates.

The space-lattice silicate, or framework silicate, results when all four oxygen atoms are shared with other tetrahedra. The tetrahedra are actually grouped to form a spiral. The sharing of oxygen atoms among the tetrahedra results in the strongest type of chemical bonds among the tetrahedra—primary valence bonding.

MECHANISMS OF LIME-SOIL STABILIZATION

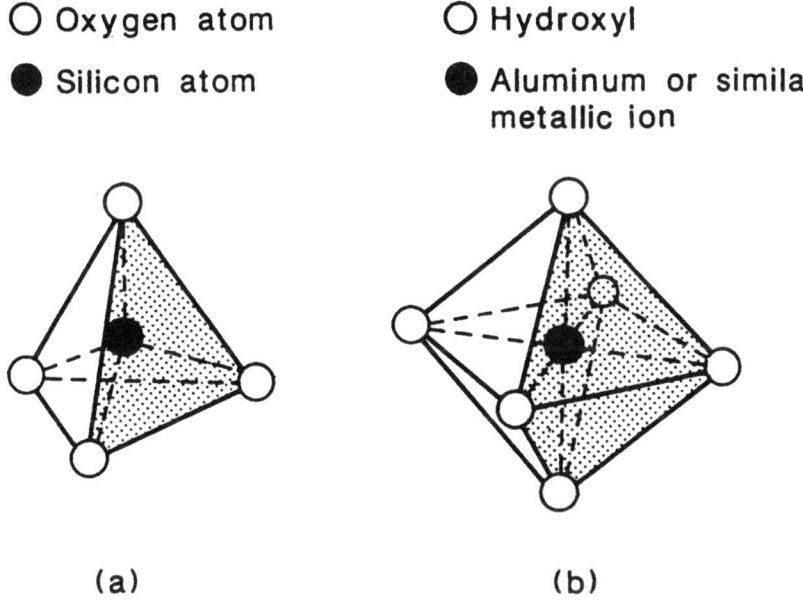

FIGURE 4.1. TWO OF THE MOST COMMON BUILDING BLOCKS OF SOILS ARE THE SILICA TETRAHEDRON (A) AND THE ALUMINUM OCTAHEDRON (B).

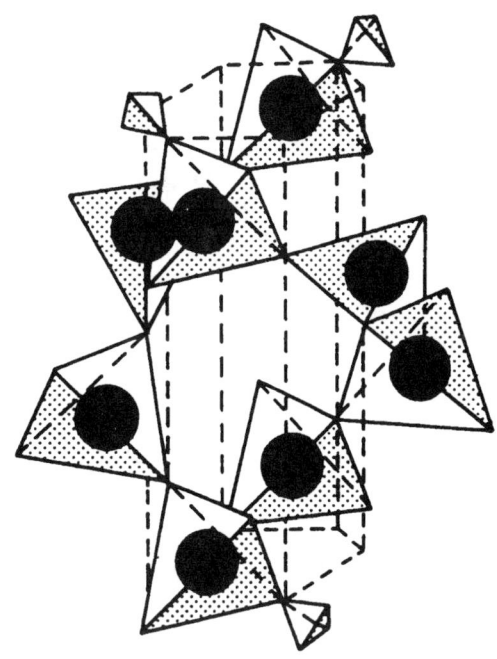

FIGURE 4.2. SILICA TETRAHEDRA IN A LATTICE ARRANGEMENT PRODUCE VERY STABLE MINERALS SUCH AS QUARTZ SAND. (AFTER MOFFATT ET AL., 1965).

A framework silicate like quartz is a very strong (mechanically) and durable (mechanically and chemically) mineral. This stability is due to the very strong internal bonding.

Another framework silicate almost identical to quartz is feldspar. In the formation of feldspar, part of the silicon positions are filled by aluminum. Because aluminum has a lower charge by +1 than silicon, a charge imbalance within the three-dimensional structure results. This charge potential is balanced by cations adsorbed within the lattice structure such as potassium, calcium and sodium. The feldspar mineral is weaker and less durable than quartz for two reasons: (1) the aluminum cation is larger than the silicon ion and does not fit among the oxygen atoms as precisely as does the silicon and (2) the adsorbed cations within the three-dimensional silicate make the structure more susceptible to weathering.

Despite the dissimilarities, both quartz and feldspar result in particles which are equi-dimensional, granular, hard and chemically relatively stable. Quartz and feldspar minerals may exist over a wide range of particle sizes from gravel-size to silt-size. In fact quartz, due to its extremely stable nature, may even retain its mineralogical structure when weathered down to clay-sized particles (less than 2 micrometers in size). On the other hand, feldspar, because it is less mechanically and chemically stable, does not retain its basic mineralogical structure upon extreme weathering. In other words, one would not generally find feldspar minerals of a clay-size.

The three-dimensional framework silicates typically compose nonplastic, granular soils.

The Clay Minerals

Silica tetrahedra are joined together only at their corners because of strong repulsion between adjacent triangles of the tetrahedra. This arrangement allows the tetrahedra to form several crystal arrangements in addition to the framework silicates. Included in these structures are independent silicates, rings or chains of tetrahedra, and sheet silicates. The sheet silicates are one unit thick but theoretically may expand infinitely in the lateral dimension. These are the units from which the clay minerals are formed.

A typical scenario for the formation of clay minerals is the chemical weathering of feldspar. Hydrolysis is probably the most important chemical weathering process and is caused by a reaction between the ions within the feldspar mineral and the dissociated hydrogen (H^+) and hydroxyl (OH^-) ions of water.

The small H^+ ions dissociated from the water molecule can easily enter the open lattice structure of the feldspar mineral. In great concentrations, they will replace metallic ions within the lattice which have been adsorbed to neutralize the charge deficiency caused by the substitution of aluminum for silicon during the formation of the feldspar. The replaced ions include sodium, potassium and calcium.

Next, the hydrogenated surfaces of the mineral become unstable, and sheets of tetrahedra and octahedral units are formed and peel off. These tetrahedral and octahedral sheets are illustrated in Figure 4.3.

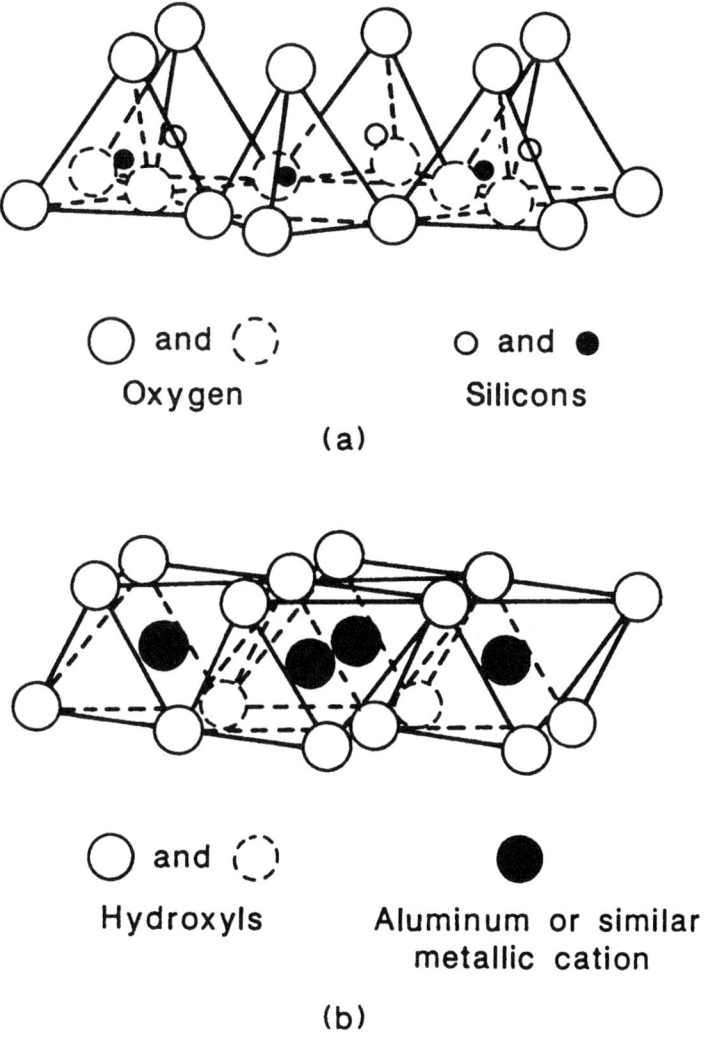

FIGURE 4.3. TETRAHEDRA AND OCTAHEDRA ARE UNITS WHICH CAN ALSO BE ARRANGED IN SHEETS WITH LARGE SURFACE AREAS.

The two basic sheets, tetrahedral and octahedral, shown in Figure 4.3 are stacked together to form the most prominent clay minerals. Two common arrangements of tetrahedral and octahedral sheets which form the basic clay mineral units are presented in Figure 4.4. The 1:1 configuration is typical of the mineral kaolinite which has low plasticity. The 2:1 configuration is typical of the clay mineral group smectite which can be very plastic and unstable.

In both configurations, a plane of atoms is common to both the tetrahedral and octahedral sheets. The bonding among the sheets is very strong primary valence bonding. However, the bonds that link the unit structures together to form particles are much weaker and are the reason for the variable response of the minerals in terms of plasticity and consistency. Before the types and characteristics of linkages between units can be discussed, the phenomena of surface charge of the unit cells must be explained.

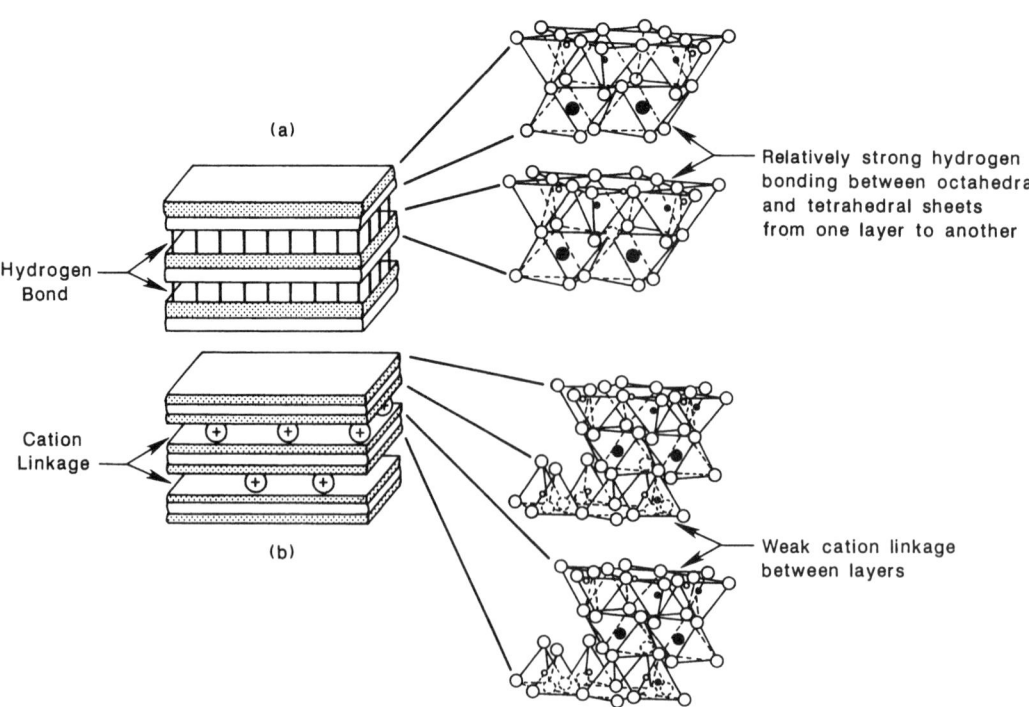

FIGURE 4.4. BASIC UNITS OF THE 1:1 MINERAL KAOLINITE (A) ARE LINKED WITH RELATIVELY STRONG HYDROGEN BONDS WHICH RETAIN A HIGH DEGREE OF MOISTURE STABILITY AMONG LAYERS WHILE THE BASIC UNITS OF THE 2:1 SMECTITE MINERAL (B) ARE LINKED BY WEAK CATION ATTRACTION. THE EFFICIENCY OF THIS LINKAGE IS A FUNCTION OF THE TYPE AND CONCENTRATION OF THE AVAILABLE CATIONS.

During the genesis of some clay minerals a phenomena called isomorphous substitution occurs. In this process some of the silicon ions in the tetrahedral sheet or aluminum ions in the octahedral sheet are replaced by other metallic ions of a lower positive charge (lower valence). The result is a charge deficiency which is reflected in a net negative charge at the surface of the clay unit cell. Some minerals, such as kaolinite, experience very little isomorphous substitution. In kaolinite, about one in every 400 silicon atoms (+4 valence), are replaced by aluminum (+3 valence) in the tetrahedral sheet. Smectite, on the other hand, experiences abundant isomorphous substitution, typically about one in six aluminum atoms (+3 valence) in the octahedral sheet are replaced by magnesium (+2 valence). The result is that smectite minerals have a very high negative surface charge and kaolinite minerals have a low negative surface charge. The ratio of surface charge between smectite and kaolinite is about 10 to 1.

The manner in which the basic unit cells of the clay minerals are linked together is strongly affected by the surface charge as well as the mineral structure. Kaolinite unit cells are linked by hydrogen bonds between oxygen atoms in the base of the tetrahedral sheet and hydroxide ions at the surface of the octahedral sheet, (Figure 4.4). Hydrogen bonding is a secondary valence bonding and is not nearly as strong as the primary valence bonds that link the sheets together. However, the bonding is strong enough to prevent the infiltration within the layers of water or foreign particles, such as cations. The result is that kaolinite is a relatively stable clay of low plasticity.

The large amount of isomorphous substitution within smectite units yields a clay mineral with a substantially negative surface charge. This charge is satisfied by the adsorption of positively charged ions, cations, at the surface. Linkage between successive layers is due to the cations which balance the surface charges. This cation linkage is a very weak bonding and results in particles of clay with prominent planes of cleavage or weakness. Due to the weak inter-unit bonding, smectite clay particles are much smaller particles than kaolinite particles.

The response of the smectite clay mineral to water is highly dependent upon the ions available in the pore water. Thus the ability to swell and shrink and demonstrate plasticity is controlled by the clay-water interaction.

Clay-water System

To this point it has been established that clay particles are generally composed of tetrahedral and octahedral sheets. In the smectite mineral, the 2:1 arrangement, Figure 4.4, coupled with a substantial level of isomorphous substitution, results in clay particles which have a plate-like morphology. These particles are also very small and possess enormous surface area. In fact, smectite minerals often have surface areas approaching 800 m^2/gm. This may be compared to only about 15 m^2/gm for kaolinite. With this enormous available surface area coupled with the highly charged nature of the clay surface, it is no wonder that the clay is so active and can so readily adsorb polar liquids like water and cations available in the environment.

Many explanations of the clay-water system are available. Some are quite complex. Figure 4.5 is an attempt to simplify the explanation while accounting for the important phenomena. First, in Figure 4.5 the negative clay surface is shown to be surrounded by positively charged cations attracted to the surface to equilibrate the charge potential. These ions are not arranged in an orderly fashion at the clay surface but form a diffused layer. This diffusion is due to like charge repulsion and thermal agitation of the cations. A second phenomenon illustrated in Figure 4.5 is the diffusion of water molecules toward the high concentration of cations. The inward diffusion is triggered by the desire of nature to move to a more random (less ordered) condition.

The water molecules not only seek to diffuse the adsorbed cation layer but are actually attracted to the cations and to the negative clay surface due to their unique dipolar structure. In Figure 4.5, the water molecules are represented as molecules with distinct positive and negative ends due to the molecular arrangement of the hydrogen and oxygen atoms. Some researchers believe that a layer of water molecules is attached to the clay surface by hydrogen bonding and that subsequent layers are more loosely

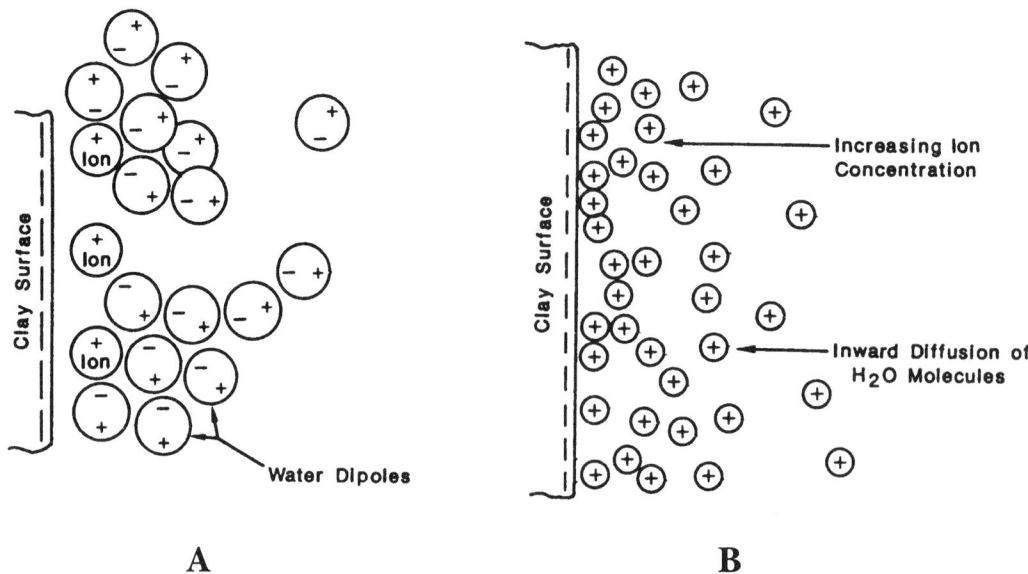

FIGURE 4.5. CATIONS AND WATER (A DIPOLAR MOLECULE) ARE ATTRACTED TO THE NEGATIVELY CHARGED CLAY SURFACE TO SATISFY THE CHARGE POTENTIAL. THIS RESULTS IN (A) ADSORBED CATIONS AND WATER MOLECULES AND (B) A DIFFUSED LAYER OF CATIONS DUE TO THEIR THERMAL ACTIVITY AND THE INFUSION OF WATER TOWARD THE CLAY SURFACE BECAUSE OF THE HIGH ELECTROLYTE CONCENTRATION. (AFTER MITCHELL, 1976).

held. At any rate, the net result is a diffused layer surrounding the clay comprised of (a) water held by hydrogen bonding and/or attracted by the diffusion gradient and (b) a diffused layer of cations, which are attracted to the negative clay surface, and anions (negatively charged ions) which are attracted to the cation and dipolar water molecules.

The net result of a highly charged surface, as is the case with smectitic clays, coupled with "unfriendly" cations, which have a single plus charge per ion and are thermally active, is a highly diffused water layer surrounding the clay particles. Because of this layer, some smectites have been documented to hold seven times their dry weight in adsorbed water. When fully hydrated these diffused water layers force the clay platelets into a parallel arrangement which offers very little shear strength.

The thickness of the diffused water layer is greatly dependent upon the type and concentration of cations available in the pore water. Divalent cations (cations with a +2 charge) can much more efficiently equilibrate the negative charge potential than can monovalent cation (cations with a +1 charge). Thus, the resulting diffused divalent cation-water layer around clay particles is much smaller than the diffused monovalent cation-water layer around clay particles of identical mineralogy.

4.04 The Dramatic Change: The Lime-Clay System

A Stable Water Layer

Lime for use in soil stabilization is most commonly produced as either hydrated high calcium lime, monohydrated dolomitic lime, calcitic quicklime or dolomitic quicklime. When lime is added to a clay-water system, the divalent calcium cations virtually always replace the cations normally adsorbed at the clay surface. This cation exchange occurs because divalent calcium cations can normally replace cations of single valence, and ions in a high concentration will replace those in a lower concentration.

The fact that calcium will replace most cations available in the water system is documented by the Lyotropic series which generally states that higher valence cations replace those of a lower valence, and larger cations replace smaller cations of the same valence. The Lyotropic series is written as:

$$Li^+ < Na^+ < H^+ < K^+ < NH_4^+ \ll Mg^{++} < Ca^{++} \ll Al^{+++}$$

where the cation to the right replaces the one to the left. Thus in equal concentrations, Ca^{++} can easily replace the cations commonly present in most clays.

Gapon (Yong and Warkentin, 1966) explains that the relationship between the cations adsorbed at the clay surface is a function of not only the concentration of cations but also the valence. Gapon's most simple and useful equation states:

$$\frac{M_e^{+m}}{N_e^{+n}} = K \frac{\sqrt[m]{M_o^{+m}}}{\sqrt[n]{N_o^{+n}}}$$

where M and N are cations of valence m and n, respectively, and e refers to exchangeable and o to ions in the pore water solution. The constant k depends on the specific cation adsorption effects and upon the clay surface. Based on the Gapon equation, equal concentrations of Ca^{++} and Na^+ ions in the solution water, natural pore water, will result in 17.5 times more Ca^{++} ions present at the clay surface than Na^+ ions. The dual effects of divalency of the calcium ion and very high concentration which would result from the addition of lime to a soil-water system are obvious.

A New Texture

The effect of exchangeable cations on the size of the diffused water layer is illustrated in Figure 4.6. Cation exchange due to the addition of lime results in stabilization of the diffused water layer and a dramatic reduction in its size. When the clay

Full Hydration

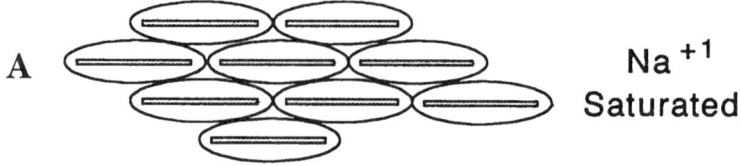

FIGURE 4.6. THE REASON FOR THE TEXTURAL CHANGE IS DUE TO THE PHENOMENON OF CATION EXCHANGE FOLLOWED BY FLOCCULATION AND AGGLOMERATION. (A) ILLUSTRATES LOW STRENGTH CLAY SOIL WHERE PARTICLES ARE SEPARATED BY LARGE WATER LAYERS. THE ADDITION OF LIME (CALCIUM) SHRINKS THE WATER LAYER (B) ALLOWING THE PLATE-LIKE PARTICLES TO FLOCCULATE.

particles are allowed to approach each other more closely due to reduction in the size of the water layer an edge-to-face attraction or flocculation occurs. Flocculation is additionally enhanced due to a high electrolyte concentration and high pH environment existing in the lime-soil-water system. The edge-to-face attraction is probably partly due to the attraction of broken bonds at the edge of the clay particles to the oppositely charged surfaces of neighboring clay particles.

The net result of cation exchange and flocculation/agglomeration of particles is:

1. Substantial reduction in size and stabilization of the adsorbed water layer,
2. Increased internal friction among the agglomerates and greater aggregate shear strength and
3. Much greater workability due to the textural change from a plastic clay to a friable, sand-like material.

Immediate Strength Improvement

Laboratory evidence substantiates textural and property changes due to cation exchange followed by flocculation/agglomeration. Table 4.1 illustrates the ability of relatively small percentages of lime to reduce the PI and swell potential of plastic, troublesome clays to innocuous levels. Figure 4.7 illustrates the *increase in shear strength,* as

Table 4.1. Atterberg Limits for Natural and Lime-Treated Soils (After Little et al., 1987).

Soil	Unified Classification	Natural Soil		3% Lime		5% Lime	
		LL	PI	LL	PI	LL	PI
Bryce B	CH	53	29	48	21	NP	
Clay Till	CL	49	27	51	12	59	11
Cowden B	CH	54	3	47	7	NP	
Drummer B	CH	54	31	44	10	NP	
Fayette C	CL	32	10		NP		
Hosmer B$_2$	CL	41	17		NP		
Piasa B	CH	55	36	48	11	NP	
Illinoian Till	CL	26	11	27	6	NP	

LL—Liquid Limit
NP—Nonplastic
PI—Plasticity Index

(Note data in Table 4.1 were provided by Dr. M. R. Thompson of the University of Illinois at Champaign-Urbana).

measured by the California Bearing Ratio (CBR) of a low plasticity clay due to textural changes in the clay. These textural changes are due to cation exchange followed by flocculation/agglomeration, and these changes essentially occur as rapidly as the lime can be intimately mixed with the clay. The soaked CBR tests (96 hour soak) in Figure 4.7 were performed immediately after compaction, without the benefits of long-term curing.

The reason for the improved shear strength is illustrated in Figure 4.6 where an unaltered clay's hydrated diffused water layer is compared to that of the same clay after lime stabilization. The ordered structure of the clay platelets surrounded by the hydrated, diffused water layers provides very little shear strength. The only resistance to relative movement is due to the overlapping and interference among the water layers. On the other hand, in the flocculated structure, the summation of the edge to face contacts provides a more substantial shear strength.

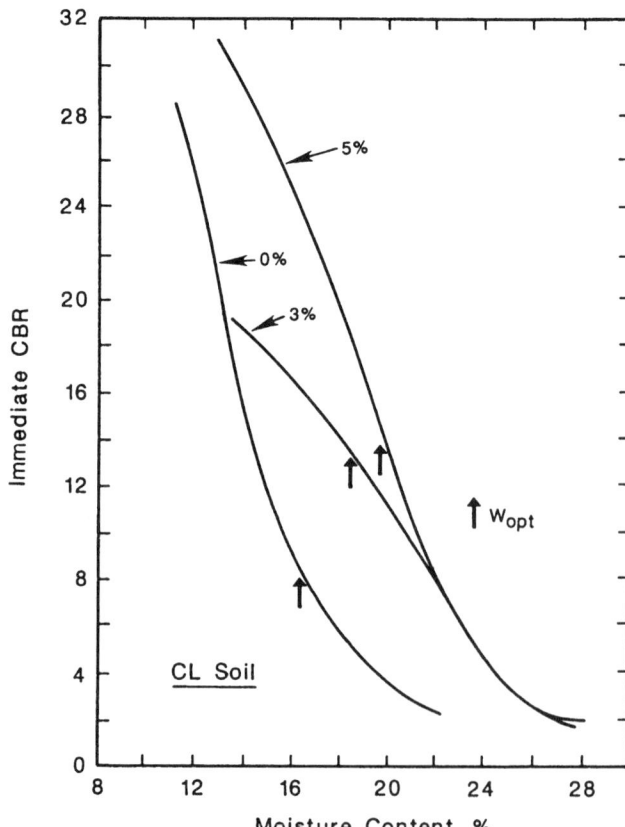

FIGURE 4.7. LIME TREATMENT PROVIDES IMMEDIATE STRENGTH GAINS AS ILLUSTRATED BY THESE CBR DATA AS A FUNCTION OF MOISTURE CONTENT FOR A CL SOIL. (AFTER THOMPSON, 1970).

Long Term Strength

The phenomenon of cation exchange and the concomitant textural changes occur with all clays. Of course, the degree of effect and/or amount of lime required to cause cation exchange is based on the chemical and mineralogical conditions of the soil and the water environment. These cation exchange and flocculation/agglomeration phenomena promote modification of the lime-soil mixture. Hence short-term physical property changes in the soil are referred to as soil modification.

Long-term strength is more complex and is heavily influenced by soil conditions and mineralogical properties. However, many clay soils are pozzolanically reactive when stabilized with lime and respond with an appreciable strength gain due to the development of a cemented matrix among the soil particles. Pozzolanic strength development which occurs with time is responsible for soil stabilization.

A pozzolan is defined as a finely divided siliceous or aluminous material which in the presence of water and calcium hydroxide will form a cemented product. The cemented products are calcium-silicate-hydrates and calcium-aluminate-hydrates. These are essentially the same hydrates that form during the hydration of Portland cement.

Clay is a pozzolan as it is a source of silica and alumina for the pozzolanic reaction. Clay-silica and clay-alumina become soluble or available in a high pH environment, (Figure 4.8). The pH of water saturated with lime is 12.45 at 25°C (77°F). Thus a soil-lime-water system has a pH high enough to solubilize silica and alumina for pozzolanic reaction. As long as enough residual calcium remains in the system to combine with the clay-silica and clay-alumina and as long as the pH remains high enough to maintain solubility, the pozzolanic reaction will continue. The reaction is illustrated by the following equations:

$Ca^{++} + OH^- +$ Soluble Clay Silica $\rightarrow$ Calcium Silicate Hydrate (CSH)

$Ca^{++} + OH^- +$ Soluble Clay Alumina $\rightarrow$ Calcium Aluminate Hydrate (CAH)

Eades and Grim (1966) skillfully adopted the pH increase phenomenon in a design procedure for lime-soil mixtures. Their procedure requires for sufficient lime to be added to the soil to satisfy all immediately occurring reactions, and yet provide enough residual lime to maintain a pH of 12.4 for sustaining the strength-producing reaction.

What is unique about the pozzolanic phenomenon is the cooperative reaction between the lime and the clay. The lime induces the high pH environment which solubilizes the silica and alumina. The lime also provides the residual free calcium which combines with the silica and alumina supplied by the clay to produce the pozzolanic reaction.

Evidence of the strength of the pozzolanic reaction comes from both field and laboratory data. Strength increases of greater than 100 psi can be achieved with many soils following 28 day curing at temperatures of approximately 21°C (70°F). Extended curing either in the laboratory or under field conditions may produce strength increases of several hundred psi. Field data indicate that with some soil-lime mixtures strength continues to increase with time up to in excess of ten years.

The Mohr-Coulomb criteria is often used to evaluate the shear strength of granular materials. This criteria states that shear strength is provided by (1) the cohesive strength of the soil and (2) the strength due to internal friction. Mathematically the law reads:

$$S = c + n \tan \phi$$

where

S = shear strength
c = cohesive strength
n = normal stress and
ϕ = the angle of internal friction.

Typical angles of shearing resistance or internal friction for lime stabilized clays are between 25° and 35°. The resulting shear strength may be as much as 100 percent higher than for the natural clay soil. This component then adds to the shear strength more and more as the confining pressure on the soil increases. The cohesion value (c)

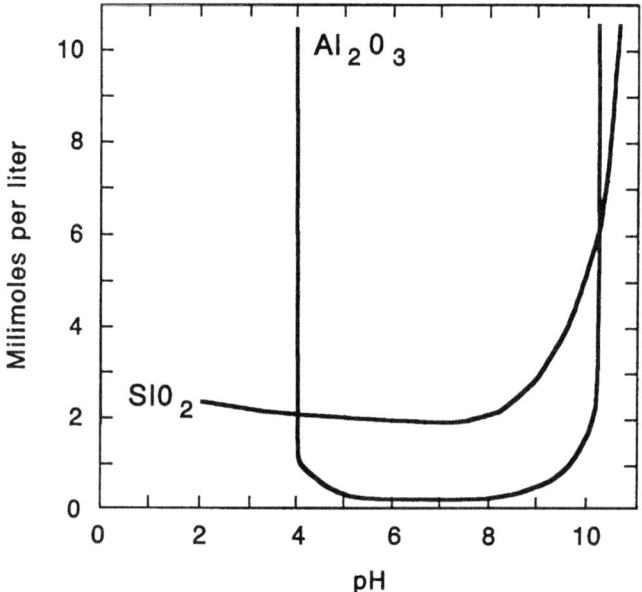

FIGURE 4.8. THE EFFECT OF A HIGH pH SYSTEM IS TO RELEASE SILICA AND ALUMINA FROM THE CLAY SURFACE. (AFTER KELLER, 1964).

increases with the compressive strength of the mixture. A rough estimate of c is 30 percent of the unconfined compressive strength (Little, et. al., 1987).

Stiffness and load spreading capability of the stabilized clay also increase with development of pozzolanic strength. Field deflection data recorded by Texas A&M University has shown that resilient moduli of plastic Texas clays in the Houston area have been increased from between 34 and 68 MPa (5,000 and 10,000 psi) for the natural clay to between 138 and 483 MPa (20,000 and 70,000 psi) after lime stabilization. Other researchers have documented stiffnesses of over 690 MPa (100,000 psi) for lime stabilized clays (TRB State of the Art No. 5, 1987).

Improvements in engineering properties of lime-soil mixtures are discussed in more detail in Chapter 6.

4.05 Evidence Supporting the Lime-Soil Reaction

Although the *exact* mechanisms of lime-soil stabilization are not totally agreed upon, general agreement does exist that four basic reactions do occur to some level:

a) cation exchange,
b) flocculation and agglomeration,
c) pozzolanic reaction and
d) carbonation.

Some researchers, such as Diamond and Kinter (1965), explain that the immediate textural changes, plasticity changes and short-term strength gains which were traditionally thought to be the result of cation exchange are actually artifacts of the crowding of calcium hydroxide molecules along the surface of the clay. This crowding results in an attack on the clay mineral surface and the formation of calcium-aluminate and calcium silicate minerals, which help bond the mineral surfaces together—reducing plasticity and affecting the textural change. This is essentially a "pozzolanic effect".

Diamond and Kinter's (1965) argument is that the surface interaction between adsorbed $Ca(OH)_2$ molecules on the clay surface which accounts for plasticity reduction and strength gain would help explain how plastic soils that are naturally calcium saturated respond to lime through plasticity reduction and other favorable changes in physical properties which occur without curing.

Diamond and Kinter (1965) attribute the rapid cementation to the immediate formation of calcium-aluminate-hydrate due to the reaction of $Ca(OH)_2$ at the edges of the clay minerals.

Eades and Grim (1965) used X-ray diffraction (XRD) and differential thermal analysis (DTA) to identify the reactions which take place between lime and clay soils. Their

XRD and DTA analysis offered mineralogical proof of the changes which occur in kaolinite, illite and montmorillonite clays when treated with lime. This mineralogical proof was supported by changes in physical properties (Atterberg limits and compressive strength) measured on the same soils. Using XRD and DTA, Eades and Grim (1965) described the reaction between lime and kaolinite clay as one in which the lime "eats into the kaolinite around the edges with a new phase forming around the core of kaolinite as a result." In contrast, reaction of lime with montmorillonite (smectite) begins with replacement of the naturally occurring cations with calcium which is provided in abundance with lime stabilization. Once enough calcium is provided to saturate interlayer positions, the clay mineral structure deteriorates, and new minerals are formed which account for strength gain and textural changes.

The findings of Eades and Grim (1965) document a very important point in the practice of lime-soil stabilization: the amount of lime necessary to initiate and "drive" lime-soil reactions which are responsible for long-term compressive strength gain and pozzolanic reactivity is soil dependent and varies considerably from soil-to-soil. This point is illustrated in Figure 4.9. These soils were cured under controlled laboratory temperature and humidity conditions.

A basic understanding of clay mineralogy and mechanisms of lime-soil reactivity provides a clearer recognition of the importance of proper lime-soil mixture design and insurance that adequate quantities of lime are added to satisfy all cation replacement and exchange reactions and provide yet enough residual lime to "drive" the pozzolanic reactions.

Little (1991) used XRD analysis to demonstrate the mineralogical change that occurs at the clay surface upon the addition of lime. The XRD spectra is used to identify clay minerals as a function of the angle of reflection of the X-ray beam. Figure 4.10 is such an XRD spectra.

In Figure 4.10a, the peak indicative of the highly expansive clay mineral smectite is a well-defined and intensive peak (at 14.6 Å for the air dry soil, 17.7 Å for the glycol solvated soil and at 10 Å for the soil dried at 550°C). Upon lime treatment, Figure 4.10b, the XRD peak indicative of smectite is diminished to approximately 25 percent of its intensity prior to lime treatment. Little (1991) found that the substantial reduction in peak intensity was due to the mineralogical changes at the clay surface due to lime-soil reaction in the high pH environment. Careful analytical measures were taken to insure that the reduction in peak intensity was due to pozzolanic-type reactivity and not solely due to carbonate surface coating nor due to change from a dispersed to a flocculated clay structure.

The significance of Figure 4.10 is that it demonstrates the mineralogical changes which result in advantageous changes in engineering properties of the stabilized soil.

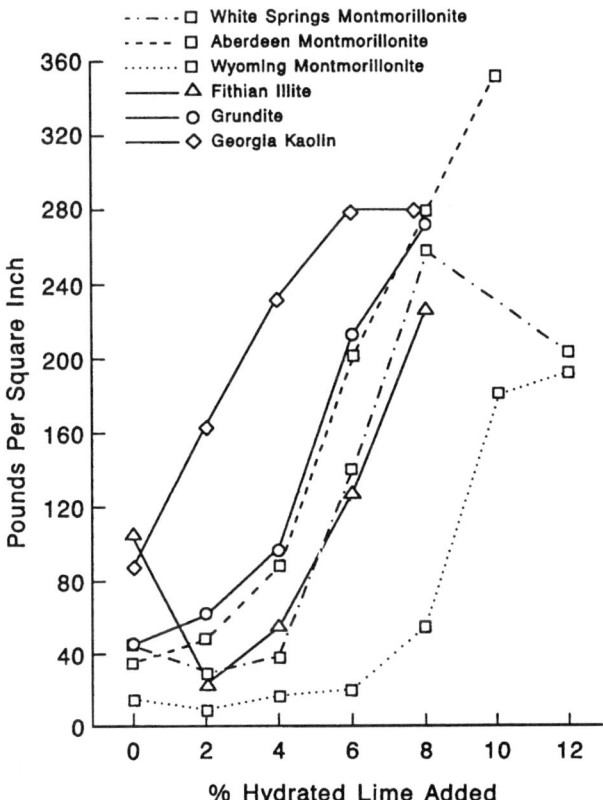

FIGURE 4.9. THE QUANTITY OF LIME REQUIRED TO PRODUCE THE POZZOLANIC REACTION (AS REFLECTED BY COMPRESSIVE STRENGTH) VARIES WITH THE TYPE AND MINERALOGY OF THE SOIL BEING STABILIZED (AFTER EADES AND GRIM, 1960).

Figures 4.11 and 4.12 provide visual evidence of this change. These figures are scanning electron microscope pictures of a Denver, Colorado, clay and an Arlington, Texas, clay, respectively, before and after lime stabilization. Note the typical morphology of the smectite clay revealed in Figures 4.11 and 4.12: thin, wavy sheets. Note the semi-crystalline to virtually amorphous product developed at the clay surface in Figures 4.11 and 4.12. This semi-crystalline to gel-like amorphous product contributes to the strength-enhancing long-term pozzolanic reaction.

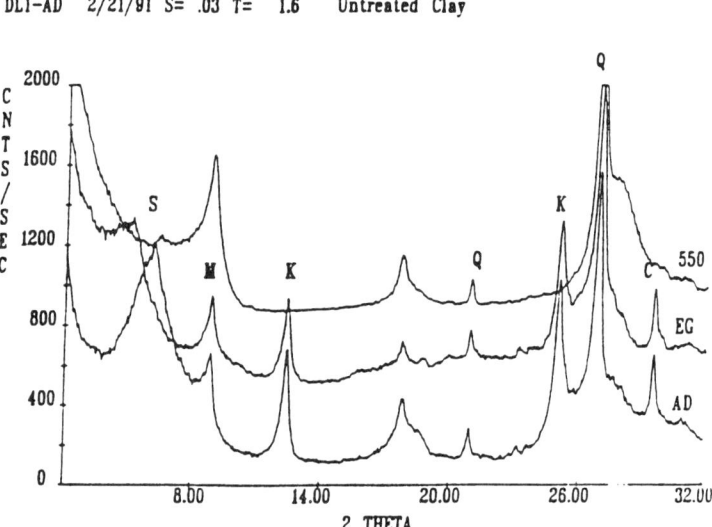

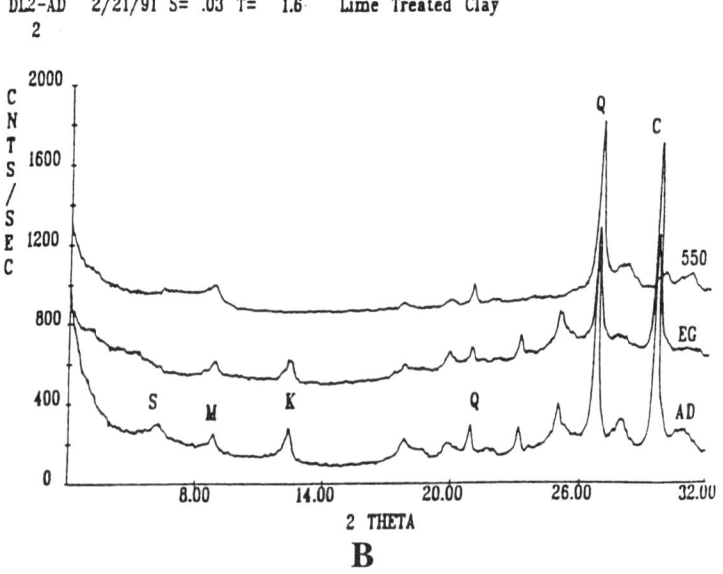

FIGURE 4.10. X-RAY DIFFRACTION (XRD) SPECTRA CAN BE USED TO PROVE THE REACTION OCCURRING BETWEEN THE LIME AND THE CLAY SURFACE. IN (A) THE XRD SPECTRA OF THE NATURAL SOIL PRODUCES AN INTENSE SMECTITE PEAKS WHILE THE PEAK ESSENTIALLY DIMENISHES UPON LIME TREATMENT (B).
NOTE: ON EACH PLOT, AD ITENTIFIES AN AIR DRY CONDITION, EG AN ETHYLENE GYLCOL SATURATED CONDITION AND 550 DRIED TO 550°C.

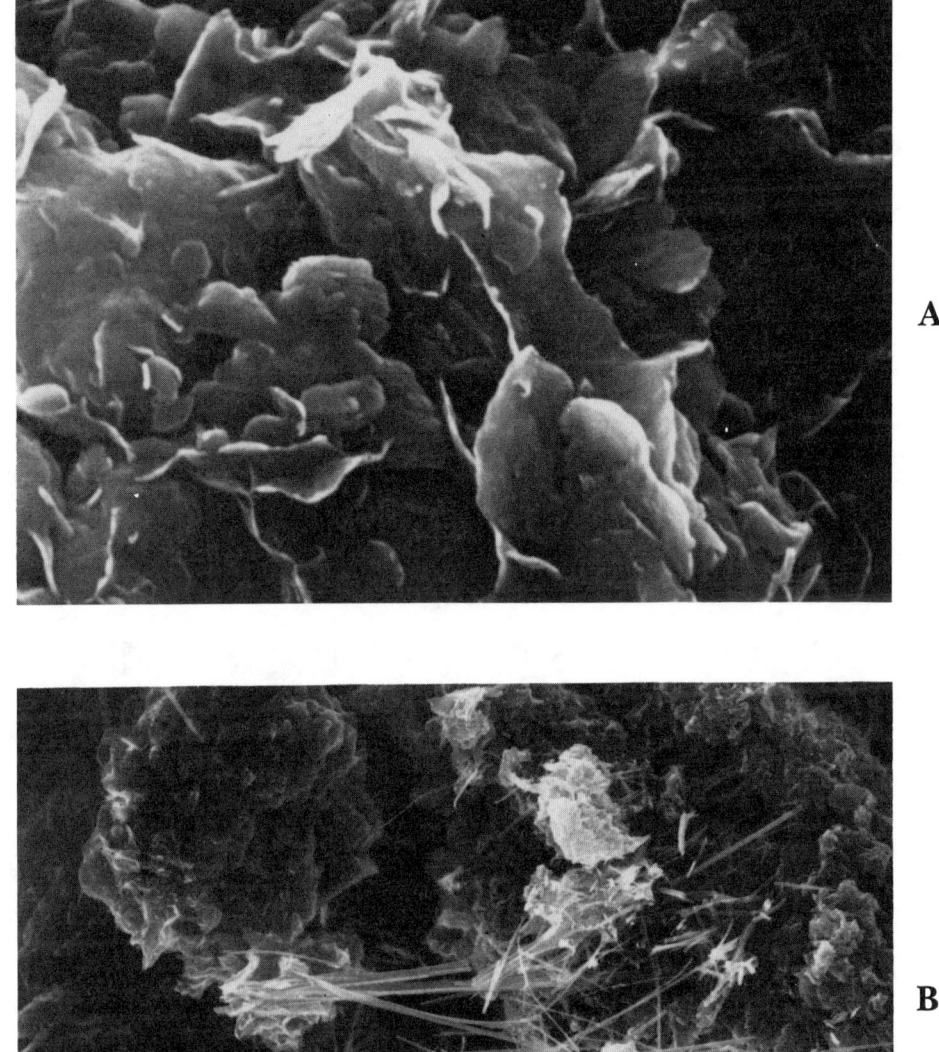

FIGURE 4.11. SCANNING ELECTRON MICROGRAPHS (SEM'S) PROVIDE VISUAL EVIDENCE OF THE DEVELOPMENT OF POZZOLANIC CRYSTALS IN THIS DENVER, COLORADO CLAY (B) THE NATURAL SOIL WITHOUT LIME IS SHOWN IN (A).

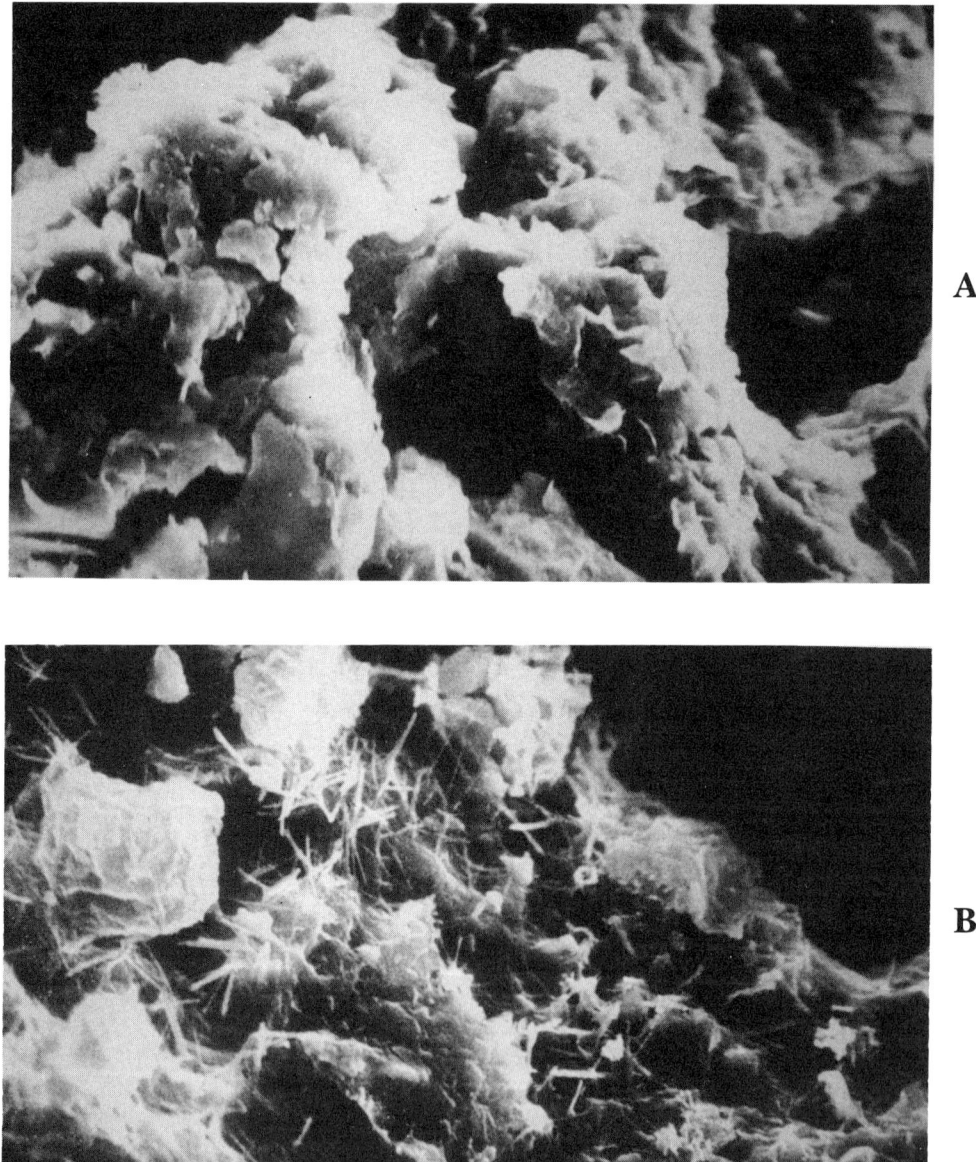

FIGURE 4.12. THE CRYSTALS PRODUCED IN SOILS VARY DEPENDING ON SOIL MINERALOGY AND TIME OF CURING. THIS ARLINGTON, TEXAS, SOIL (A) DEMONSTRATES CONSIDERABLE POZZOLANIC PRODUCT DEVELOPMENT UPON ADDITION OF LIME (B).

4.06 Suitable Soils for Lime Stabilization

Experience has shown that lime will react with medium, moderately fine and fine-grained soils to produce decreased plasticity, increased workability, reduced swell and increased strength. Generally speaking, those soils classified by the Unified System as CH, CL, MH, SC, SM, GC, SW-SC, SP-SC, SM-SC, GP-GC or GM-GC are potentially capable of being stabilized with lime.

The key to a pozzolanic reaction resulting in long-term strength gain is the presence of a reactive clay to provide the pozzolans. Although lime cannot react pozzolanically with sands (composed of framework silicates) which have no clay fraction, lime may be an effective stabilizer with sandy or silty soils which have a clay content as low as seven percent and a plasticity index as low as 10 (Little, et. al., 1987).

As a general guide to stabilization Little, et. al., (1987) proposed that lime stabilization should be *considered* as a stabilizer or as a pre-stabilizer for soils having plasticity indices of greater than 10 and with greater than 25 percent of the soil smaller than the number 200 sieve. In the case of plasticity indices above 30 and greater than 25 percent material passing the number 200 sieve, the selection criteria recommended by Little, et. al. (1987) for use by the Air Force recommends the use of lime either as the stabilizer of choice or as a pre-treatment to reduce the plasticity index below 30 followed by Portland cement stabilization. See Figure 4.13.

The extent to which the soil-lime pozzolanic reaction proceeds is influenced primarily by natural soil properties. With some soils, the pozzolanic reaction is inhibited, and cementing agents are not extensively formed. Those soils that react with lime to produce substantial strength increase (greater than 345 kPa (50 psi) following 28 day curing at 22°C (73°F) are "reactive" and those that display limited pozzolanic reactivity (less than 345 kPa (50 psi) strength increase) are "nonreactive" (Thompson, 1970).

4.07 Factors Which Influence Lime-Soil Interactions

A number of factors influence lime-soil pozzolanic reactions. These factors include: degree of weathering, soil-water pH, base cation concentrations, silica-alumina concentrations, organic content of the soil and soluble sulfate content of the soil. The first four factors influence the rate and the degree of the pozzolanic reaction. The method in which this occurs is controversial and varies considerably among different soil types and different climatic regions. However, some general trends regarding the effects of these factors are discussed in the following paragraphs. The factors of major consequence in terms of their influence on pozzolanic reactivity are discussed first: organic carbon and sulfates. The other factors affecting pozzolanic reactivity are discussed under the headings of Clay Content, Nature of Clay (Mineralogy), Weathering, Pedology and Geographical and Climatic Effects.

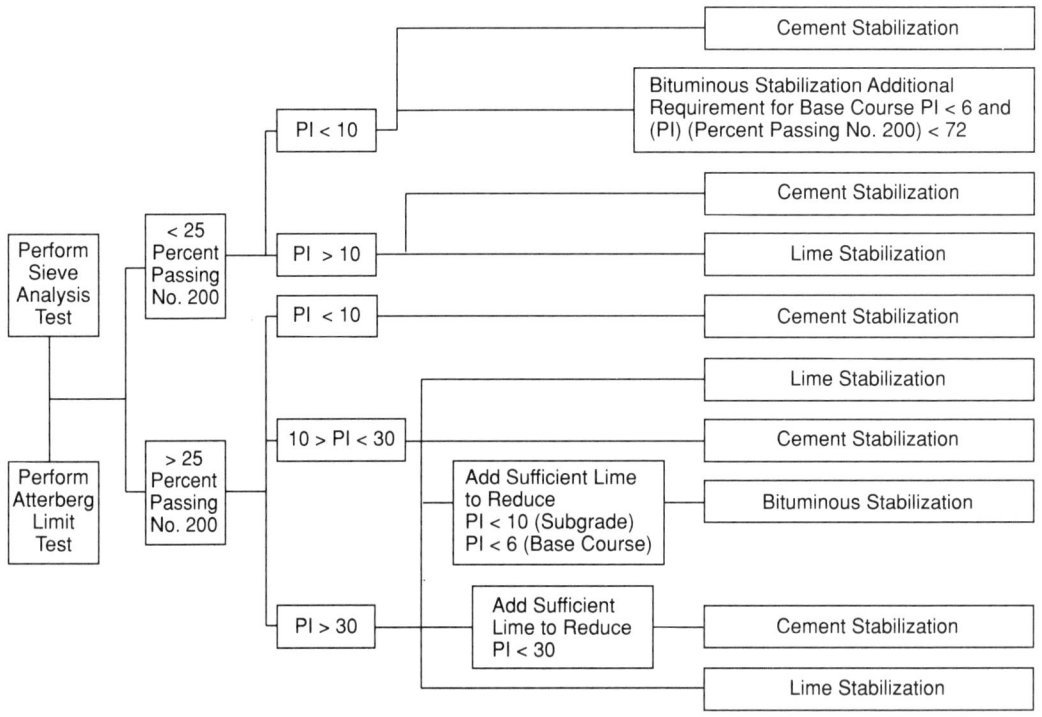

FIGURE 4.13. A GOOD ESTIMATE OF THE APPROPRIATE STABILIZER FOR A CERTAIN SOIL CAN BE DETERMINED BASED ON SIMPLE AND EASY TO MEASURE SOIL INDEX PROPERTIES SUCH AS MINUS 200 CONTENT AND PI (AFTER CURRIN ET AL., 1976).

Organic Carbon

Organic carbon can inhibit the reaction between calcium and the clay mineral surface. This occurs generally because the organic molecule is quite complex and can adsorb calcium cations or interact with soil exchange sites and hence prevent them from reacting with the soil as they normally would to produce cation exchange and pozzolanic reaction. It is difficult to predict exactly what level of organic material is enough to substantially interfere with soil-lime reactivity. This is influenced by the type of soil being stabilized and the nature of the organic material.

As a general rule an organic content in excess of one percent is cause for concern that the organic material will interfere with the pozzolanic reaction. The solution to this problem will range from removing the soil because it cannot be effectively stabilized with a calcium-based stabilizer to simply adding additional lime. The questions which arise with regard to how high organic carbon contents affect lime-soil reactivity should be answered in the mixture design process. This will be discussed in Chapter 5.

The Effect of Sulfates on Lime-Soil Interactions

Soluble sulfates present a concern when stabilizing the soil with any calcium-based stabilizer. The basic reason for this is relatively simple: soluble sulfates in high enough concentrations can interfere with pozzolonic reactions. The culprit is a reaction product that consists of calcium, alumina, water and sulfate. This calcium-sulfate-aluminate-hydrate can occur in two forms: a high sulfate form, ettringite, and a low sulfate form, monosulfoaluminate. Either form can be harmful as can the potential to transition between forms.

The reason that the calcium-sulfate-aluminate-hydrates are harmful is because of the expansion that results from their formation and because this expansion can occur with a high potential pressure (approximately 241 MPa (35,000 psi) within the crystal). This is different from the traditional pozzolanic reaction products such as calcium silicate hydrates (CSH) and calcium aluminate hydrates (CAH). These products stop growing when they reach an obstruction, such as soil, aggregate or other pozzolanic product.

Calcium-sulfate-aluminate-hydrate is not a problem if it forms prior to compaction of the soil. However, when it forms after compaction or even after the roadway has been placed in service, the resulting heave can be very destructive.

Recent research by Little (1992), and Petry (1992) demonstrates that the use of modified construction techniques can, under certain circumstances, avert the harmful reactions of sulfate-induced heave. Generally, these construction techniques stress basic points of good lime-soil construction:

(a) That plenty of water is available to solubilize the sulfates and initiate formation of the calcium-sulfate-aluminate-hydrate products during the mixing and the mellowing period and prior to compaction,
(b) That all soil, lime, sulfates and water are homogeneously mixed so that if calcium-sulfate-aluminate-hydrate products have the potential to form the majority of them will do so in a controlled manner prior to compaction and
(c) The pavement structure is designed to provide proper drainage so that high sulfate water is prevented from permeating the stabilized layer-leading to potentially deleterious post-stabilization reaction.

Since the formation of calcium-aluminate-sulfate-hydrates occurs rapidly under appropriate conditions, inducing their reaction early-on in the construction process and prior to compaction is the desired process. The worst case scenario is when the reaction potential exists in a clay (containing a high percentage of soluble sulfates stabilized with a calcium-based stabilizer) but the reaction does not occur until the pavement is placed in service. An example is if quicklime is used to stabilize a high soluble sulfate content clay but the quicklime is poorly distributed and mixed with insufficient water. The reactants could lie "dormant" until water carries soluble sulfates from the soil to the improperly mixed quicklime. The water will then hydrate the lime, increase the pH of the clay which will release alumina in the high sulfate environment. The result could be the formation of calcium-aluminate-sulfate-hydrate. Proper mixing and moisture control *may* be able to prevent this condition from occurring or minimize the effects.

On-going research may be able to provide answers that will help solve the problems with stabilizing high clay content, high sulfate content soils. The history of lime stabilization in the State of Texas is documentation that lime can be successfully used to stabilize sulfate-bearing soils as indeed lime has successfully stabilized clay soils containing significant levels of sulfates. However, until on-going research provides more explicit guidelines, caution must be used when using lime or any calcium-based stabilizer with soils containing more than 10 percent clay and 0.2 percent soluble sulfates when the sulfates are extracted from the soil in a 10 parts water to 1 part soil solution. More detailed information on the potential effects of sulfates and what can be done to prevent these reactions can be gained from Petry and Little (1992) and Little and Petry (1992).

Clay Content

As previously discussed, lime-soil pozzolanic reactions occur between soil silica and/or soil alumina and lime to form cementing agents identified as calcium silicate hydrates and calcium aluminate hydrates. The clay minerals, quartz, feldspar, micas and other silicates or aluminosilicate minerals in the soil are possible sources of silica and alumina. The addition of lime serves the dual purpose of freeing silicates by increasing the pH and supplying the calcium divalent cation. Since the clay minerals are the most abundant source of readily available silica and alumina, an adequate clay content must be present to provide a source for the pozzolanic reaction.

Nature of Clay (Mineralogy)

Generally, the most pozzolanically reactive clay minerals are the montmorillonitic and mixed layered clay minerals. This is probably due to the greater specific surfaces of smectite and mixed layer materials with increased availability of silicates and aluminates. These minerals are more freely attacked and dissolved in the lime-induced high pH environment. A conclusion relating clay mineralogy and silica and alumina availability to pozzolanic reactivity must, however, be tempered with the realization that other environmental factors may override.

It is important to remember that even though the smectite and mixed layer clay minerals may be more reactive, other clay minerals still frequently exhibit high reactivity.

Weathering

Thompson's work with Midwestern soils (1966) reveals the validity of pH, weathering and soil drainage as indicators of lime-soil reactivity. Generally, high natural pH's (above 7) indicated good lime reactivity. Soil pH's below 7 indicated poorer reactivity.

In poorly drained soils, breakdown and removal of soil constituents are retarded and leaching effects are minimized (Thompson, 1966) maintaining a high pH level. Relatively unweathered soil minerals in poorly-drained soils are a readily available silica and/or alumina, or pozzolan, source to react with lime. Highly weathered soils indicated by low soil pH should be less reactive.

The degree of weathering and type of natural soil drainage influence certain identifying factors as to a soil's lime reactivity. The Ca/Mg ratio is indicative of the degree of weathering. As a soil is weathered, exchangeable Ca^{++} is leached and Mg^{++} remains in the exchange complex. Therefore, the Ca/Mg ratio is reduced with weathering. Percentage base saturation is another weathering indicator for temperate soils. As metallic cations are leached from the soils by weathering, their positions are filled in the exchange complex by H^+ cations. As a larger number of H^+ ions are adsorbed, percentage base exchange saturation and pH decrease. Therefore, a low base saturation means a high degree of weathering. The percentage base saturation and Ca/Mg ratio demonstrates the importance of soil weathering status on lime reactivity. In slightly to moderately weathered soils, the degree of depletion of soil constituents is small, and has little effect on the soil's lime reactivity (Thompson, 1966). Still present are constituents not highly resistant to weathering. These serve as excellent sources of silica and/or alumina. Base saturation percentage and the Ca/Mg ratio seem to be good indicators of lime-soil reactivity in soils of the temperate climatic regions.

In poorly drained soils iron usually exists in the ferrous, Fe^{++}, state whereas in better drained soils, it exists in the ferric, Fe^{+++}, state. Iron oxides in poorly drained soils are generally not well distributed and exist in concretions. In the better drained soils, they are more colloidal and are more uniformly distributed throughout the soil's B horizon. Joffe (1947) believes that iron oxides may coat soil minerals. Therefore, the well drained colloidal Fe_2O_3 could coat the clay mineral surfaces retarding pozzolanic reaction.

Pedology

Since soil properties are very greatly influenced by their relative location within the soil profile, one should readily expect a correlation between soil horizons and lime reactivity.

Lime reactivity in the A horizon may be affected by organic matter content, weathering and subsequent loss of soil constituents. B horizon soils generally display substantial reactivity (Thompson, 1966), although some do not. Lack of B horizon reactivity may be related to natural drainage, pH, percentage of base saturation or Ca/Mg ratio (Thompson 1966).

Thompson (1966) points out that the high level of reactivity of B horizon soils are probably due to the accumulation of weathering products and fine clay from the upper part of the profile. This fine clay is highly susceptible to dissolution by high pH environments produced by lime and will serve as a good silica and alumina source. The C horizons are not normally subjected to weathering and should be excellent silica and alumina sources for substantial lime-soil pozzolanic reaction. However, C horizons have no accumulation of additional material and may not be as good a source of readily dissolved silica and alumina as the B horizon soils.

Geological and Climatic Effects

The fact that lime reactivity indicators vary with soil environment and geographical location is evidenced by Hardy's work with tropical and subtropical soils (1970). No single soil property proved to be an accurate predictor of lime reactivity for these soils. Two or more soil properties or characteristics are required.

Hardy's work (1970) showed that organic carbon in excess of one percent hindered stabilization. Soil pH, cation exchange capacity and percent base saturation are useful indices of lime reactivities within tropical and subtropical ultisols. High pH and base saturation values indicated good ultisol reactivity. Low cation exchange capacity (CEC) indicated poor ultisol reactivity.

Probably the best index of lime reactivity for tropical and subtropical oxisols is the relative concentration of the basic soil constituents as measured by the silica sesquioxide (where a sesquioxide is a compound in which two metal cations are combined with three oxygen atoms) ratio and to a lesser extent the silica-alumina ratio. Low silica sesquioxide and silica alumina ratios indicate good oxisol-lime reactivity. Lime requirements to maximize the strength of lime-treated tropical and subtropical soils are generally higher than those of temperate zone soils (Hardy, 1970). It is evident that the type and degree of the weathering process which has predominated in a soil profile significantly influences the state of basic soil constituents and hence is the primary influencer of a soil's lime reactivity. Furthermore, the value of lime-soil reactivity indicators such as profile drainage, extractable iron, the presence of free carbonates and the presence of sulfates vary significantly between temperate and tropical and subtropical soils.

4.08 Important Sources of Information Concerning Soil-Lime Reactivity Potential

The single most important source of information about soils and their potential for stabilization is the U. S. Department of Agriculture's County Soil Report. These reports are available for approximately 90 percent of the counties in the United States. Their original and prime purpose is to provide important information about the use of soils for agricultural purposes. However, almost all county soil reports contain a wealth of vital information concerning the engineering uses of the soils.

The valuable information about soils that can be readily obtained from the County Soil Report and used for decisions concerning the use of these soils in roadways or for other earth construction and concerning the stabilization of these soils includes:

1. Summary of the basic geology of the area,
2. Discussion of the soil development or pedology of the area and each soil type,
3. Detailed descriptions of the soil profiles of each soil association in the county,
4. Typical gradations, Atterberg limits, soil classifications (typically Unified and AASHTO), pH values, permeabilities, cation concentrations, drainage characteristics and mineralogical descriptions of the various soil types in the county and
5. Aerial photographic maps of the county with soil types superimposed on the map to aid in location of soil types for a given project.

The value of the County Soil Report to determine the potential for the lime stabilization of the soil is evident. Figure 4.14 illustrates the type of information available in a county soil report describing the soil profile and the engineering properties of soil series, respectively.

Example:	Houston Black: HnA, HnB, HnC2, HoD2, HsD
Road Subgrade:	Poor: High shrink-swell potential; poor traffic-supporting capacity
Highway Location:	Severe: high shrink-swell potential; poor traffic-supporting capacity
Example:	Houston Black Clay: 1 mile north of Manor on a paved country road, then 50 feet east into a cultivated field (Model)
Liquid Limit:	A 67 B 69 C 68
Plasticity Index:	A 41 B 46 C 49
Classification: AASHO:	A A-7-6(20 B A-7-6(20) C A-7-6(20)
Unified[2]:	A Ch B Ch C CH

FIGURE 4.14. ILLUSTRATION OF INFORMATION AVAILABLE IN COUNTY SOIL REPORT (TRAVIS COUNTY, TEXAS).

4.09 References

Currin, D. D., Allen, J. J. and Little, D. N., (1976). "Validation of Soil Stabilization Index System with Manual Development," Report FJSRL-TR-76-0006, Frank J. Seiler Research Laboratory, United States Air Force Academy, Colorado.

Diamond, S. and Kinter, E. B., (1965). "Mechanisms of Soil-Lime Stabilization, An Interpretive Review," *Highway Research Record* No. 92.

Eades, J. L. and Grim, R. E., (1960). "Reactions of Hydrated Lime with Pure Clay Minerals," *Highway Research Board Bulletin* No. 262.

Eades, J. L., (1962). "Reactions of $Ca(OH)_2$ with Clay Minerals in Soil Stabilization," Ph.D. Dissertation, University of Illinois—Champaign—Urbana.

Eades, J. L., Nichols, F. P. and Grim, R. E., (1962). "Formation of New Minerals with Lime Stabilization as Proven by Field Experiments in Virginia," *Highway Research Board Bulletin* No. 335.

Hardy, J. R., (1970). "Factors Influencing the Lime Reactivity of Tropically and Subtropically Weathered Soils," Ph.D. Dissertation, University of Illinois—Champaign—Urbana.

Joffee, J. S., (1949). "Pedology," Pedology Publications, New Brunswick, New Jersey.

Keller, W. D., (1964). "Processes of Origin and Alteration of Clay Mineral," *Soil Clay Mineralogy*, C. I. Rich and G. W. Kunze, Editors, University of North Carolina.

Little, D. N., Thompson, M. R., Terrel, R. L., Epps, J. A. and Barenberg, E. J., (1987). "Soil Stabilization for Roadways and Airfields," Report ESL-TR-86-19, Air Force Engineering and Services Center, Tyndall Air Force Base, Florida.

Little, D. N., (1991). "X-Ray Diffraction, Energy Dispersive Spectra and Scanning Electron Microscopic Evaluation of Stabilized Denver Clay Soils Containing Sulfates," Report to Chemical Lime Company, Fort Worth, Texas.

Little, D. N. and Petry, T. M., (1992). "Recent Developments in Sulfate-Induced Heave in Treated Expansive Clays," Second Interagency Symposium on Stabilization of Soils and Other Materials, Metarie, Louisiana.

Mitchell, J. K., (1976). *Fundamentals of Soil Behavior*, John Wiley and Sons, Inc.

Moffatt, W. G., Pearsall, G. W., and Walff, J., (1965). *The Structure and Properties of Materials*, Volume 1, Structure, John Wiley and Sons, Inc. New York.

Petry, T. M. and Little, D. N., (1993). "Evaluation of Problematic Sulfate Levels Causing Sulfate Induced Heave in Lime-Stabilized Clay Soils," *Transportation Research Record*.

Thompson, M. R., (1966). "Lime Reactivity of Illinois Soils," *Journal of Soil Mechanics and Foundation*, ASCE, Vol. 92, No. SMS.

Thompson, M. R., (1967). "Factors Influencing the Plasticity and Strength of Lime-Soil Mixtures," Bulletin 492, Engineering Experiment Station, University of Illinois—Champaign—Urbana.

Thompson, M. R., (1970). "Suggested Method of Mixture Design Procedures for Lime-Treated Soils," American Society of Testing and Materials, Special Technical Publication No. 479.

Transportation Research Board, (1987). *State of the Art Report No. 5, Lime Stabilization.*

Yong, R. N. and Warkentin, B. P., (1966). *Introduction to Soil Behavior,* MacMillian, New York.

CHAPTER 5

MIXTURE DESIGN

5.01 Objective of Mixture Design

The primary objective of mixture design is to identify an optimum lime content to be used during construction to modify or stabilize the soil or aggregate. The optimum lime content is a function of the expectations of how the stabilized material will be used. This is because a fairly wide range of lime contents can be used based on the desired engineering properties of the lime-soil/aggregate mixture. The desired objectives may range from a reduction in plasticity and construction expediency (modification) to permanent engineering changes which affect the strength of the mixture and performance of the pavement which contains the treated layer (stabilization).

5.02 Mixture Design Criteria

Modification

Mixture design can be divided into two categories. In the first category, the objectives are soil-modification plasticity reduction, improved workability, immediate shear strength increase and reduced volume change potential. These characteristics are generally associated with rapidly occurring reactions between lime and the soil. Cation exchange and the associated phenomenon of flocculation and agglomeration occur almost simultaneously with intimate mixing of lime with the soil during the construction process and are often given credit for the type of property and textural changes associated with this category of objectives. However, some rapidly occurring pozzolanic-type reactions must also contribute to these changes. Mixture design criteria typically associated with this category of stabilization include (TRB Lime Stabilization State of the Art Report (1987)):

1. No further decrease in PI with increased percentage of lime,
2. Acceptable PI reduction for the particular stabilization objective,
3. Acceptable swell potential reduction and
4. Shear Strength (i.e., CBR or R-value) increase sufficient for anticipated uses.

Actual values for the above listed criteria are usually established relative to the properties of the natural soils being stabilized.

Stabilization

The second category of criteria involves strength related criteria which are primarily associated with the pozzolanic reaction between the lime and the soil minerals which comprise the soils and aggregates. Strength criteria can be rationally related to pavement performance and pavement structural design. Additionally, lime-soil mixture strengths are associated with durability. Strength and durability are objectives of stabilization.

Since base course materials are located closer to the surface and usually encounter higher load-induced stresses than subbases, higher strengths are required of base layers than subbase layers. Additionally, base courses normally must meet a higher standard of reliability and durability than subbases. And since durability can be improved with higher strengths and variability can be minimized with increased strengths, it is reasonable that base courses have more strict strength requirements than subbase layers.

In pavement design, compressive stresses, tensile stresses, shear stresses and flexural tensile stresses are associated with different modes of pavement distress and are all important considerations. Permanent deformation in the pavement is associated with compressive and shear stresses. Tensile failure and flexural fatigue failure are associated with tensile strength and flexural tensile strength of the mixtures. Fortunately, one test, the unconfined compressive strength test, provides all the information necessary about the mixture for mixture design. This is because the tensile and flexural strengths of these mixtures can be reliably predicted from the unconfined compressive strengths.

5.03 Current Mixture Design Procedures

A large number of mixture design procedures exist. This is primarily because various states and agencies have developed particular criteria and procedures to fit their specific design needs and lime-soil mixture property objectives. Mixture design criteria have been validated on the basis of field performance in various states and by various agencies. Since various mixture design procedures are developed to meet specific user objectives and are validated for specific geographic regions, indiscriminate application of mixture design procedures to areas other than for which they were designed is discouraged.

Several mixture design procedures are summarized in a number of publications. The TRB State of the Art Report 5 (1987) summarizes the California procedure, Eades and Grim procedure, Illinois procedure, Oklahoma procedure, South Dakota procedure, Thompson procedure and Virginia procedures. Three procedures have been selected for presentation in this handbook as they illustrate three sound approaches to mixtures design. These are the Thompson procedure, the Eades and Grim procedure and the Texas procedure. The TRB Lime Stabilization State of the Art Report 5 (1987) was used extensively as a reference in summarizing these procedures.

Thompson Procedure

Treatment Level

Most fine-grained soils can be effectively stabilized with 3 to 10 percent lime (on a dry weight basis of soil). Under normal field conditions, approximately 2 percent lime is the minimum quantity that can be effectively distributed and mixed with a fine-grained soil.

Mixture Design Protocol

The basic components of mixture design are:

1. Method of preparing the soil-lime mixture,
2. Procedures for compacting and curing specimens,
3. Testing procedures for evaluating a selected property or properties of the soil-lime mixture and
4. Appropriate criteria for establishing the design lime content.

Mixture Preparation

Lime content is specified as a percentage of the dry weight of soil. Soil-lime mixtures are prepared by dry mixing the proper amount of soil and lime and blending the required amount of water into the mixture. ASTM D-3551 should be followed. The mixture should be allowed to mellow for approximately one-hour prior to specimen preparation. Mixtures are normally prepared at or near optimum moisture content as determined by ASTM D-698 or D-1557. Other moisture contents may also be used. In some situations a moisture content may be selected to represent an in situ field condition (TRB Report 5, 1987).

Density Control

The density of the compacted specimens must be carefully controlled. The strength of a cured soil-lime mixture is greatly influenced by density and small density variations make it difficult to accurately evaluate the effect of other variables such as lime percentages and curing conditions. Thus, the compactive effort should always be specified. ASTM D-698 or equivalent density is recommended for normal mixture design purposes. Other compactive efforts may be used to simulate anticipated field conditions (TRB Report 5, 1987).

Curing Conditions

Time, temperature and moisture must be controlled. For stabilization applications where "immediate" strength is an important factor, specimens can be tested immediately after compaction. Ambient temperature or accelerated (high temperature) curing are used for applications where field curing can be achieved prior to use for the stabilized layer.

Laboratory curing conditions should be correlated with field conditions. Because the first winter's exposure is most critical, for freeze-thaw zones, it is important to approximate the "field strength" of the mixtures before the beginning of the winter.

Normal curing conditions are 21°C (72°F) for 28 days. Accelerated curing conditions are 49°C (120°F) for 48 hours (Little et al. (1987)).

Specimens should be cured in a "sealed container" to prevent moisture loss and lime carbonation. Sealed metal cans, plastic bags, etc. are satisfactory.

Disparities in curing conditions make it difficult to compare the results obtained from different testing methods. Mixture quality criteria developed for a particular test procedure should not be arbitrarily adopted for analyzing test results obtained from a different test method.

Testing Procedures

Moisture-density relations, plasticity characteristics, swell potential, uncured strength and cured strength are significant soil-lime mixture properties. Recommended testing procedures are presented below.

   a. *Moisture-Density Relations.* Utilize ASTM D-698. In many instances lime stabilization is used under conditions (wet soils, poor support, etc.) where it may be very difficult to achieve a high percentage of specification density, but adequate soil-lime mixture properties are obtained at lower densities.
   b. *Atterberg Limits Procedure.* ASTM D-4318 should be used to determine the plasticity characteristics of the soil-lime mixture. The mixture should not be cured prior to determining the PI since the field objective is related to obtaining immediate improvement and substantial pozzolanic strength development is not required.
   c. *Swell Potential.* Use ASTM D-3668 to evaluate swell potential.
   d. *CBR Test.* The CBR test is appropriate for the following conditions:
      (1) "Immediate" (uncured) strength is a major factor. In this situation, the soil-lime mixture is not highly cemented.
      (2) The soil-lime mixture does not gain significant cured strength due to limited soil-lime-pozzolanic cementing reactions, and the mixture is considered a "modified" soil.

Conduct the CBR test in accordance with ASTM D-3668. The specimens may be either soaked or unsoaked depending on the stabilization objective. Unsoaked conditions may be appropriate for some "immediate strength" evaluation purposes.

For expedient, comparative testing procedures, CBR penetration tests (as per ASTM D-3668) can be conducted on "Proctor-sized" (101.6-mm diameter by 114.3-mm, 4-inch diameter by 4.6-inch) specimens prepared in the process of determining the moisture-density relation of a soil-lime mixture. The data provide comprehensive moisture-density and "immediate CBR" information for the soil-lime mixture.

    e. *Unconfined Compression Test.* Unconfined compression test (ASTM D-5102) procedures should be used to evaluate soil-lime mixtures which develop significant cured strength. A strength gain of 345 kPa (50 psi) cured (28 days at 21°C (72°F) or equivalent) soil-lime mixture strength minus strength of natural soil] indicates that the soil-lime pozzolanic cementing reaction is proceeding.

Compressive strength testing should be performed in accordance with the procedure presented in Appendix 5.01 or in accordance with ASTM D-5102. Fifty-one-mm (2.0-inch) diameter by 101.6-mm (4.0-inch) high specimens are recommended. Since the length to diameter ratios (l/d ratios) vary among test methods, compressive strength values should be corrected to an l/d ratio of 2 for comparison and specification purposes.

Mixture Design Criteria

Mixture design criteria are used to evaluate the adequacy of a given soil-lime mixture. Criteria vary depending on the stabilization objectives and anticipated field service conditions, i.e., environmental factors, wheel loading considerations, design life, etc. Mixture design criteria may, thus, range over a broad scale and are based on careful considerations of the specific conditions associated with the stabilization project. For soil-lime mixtures used in structural layer applications, minimum strength requirements are specified. Design lime content is normally that percentage which produces maximum strength for given curing conditions.

Strength criteria are specified in terms of compressive strength. Minimum strength requirements are higher for base materials than for subbase materials since stress and durability conditions differ for various depths in the pavement structure.

Cured compressive strength criteria for various structural layer applications are presented in Table 5.1.

Lime modification is used to expedite construction (improve workability, facilitate drying and form a "working platform") or to modify the in situ subgrade or embankment soil properties (increase CBR, decrease swell potential, decrease plasticity).

For construction expedient and subgrade modification purposes, design lime content can be based on an evaluation of the effect of lime content on the "uncured" CBR

Table 5.1 Cured Strength (28 Days) Requirements For Soil-Lime Structural Layers (Modified from Thompson, 1970).

Layer Type	No Freeze-Thaw Activity	Freeze-Thaw* Zone
Base	1,034 kPa (150 psi)	1,379 kPa (200 psi)
Subbase	689 kPa (100 psi)	1,034 kPa (150 psi)

* Use these criteria if F-T cycles will occur in the structural layer. It is possible to be in a mild F-T area and *not* experience F-T cycles in the subbase or base layer.

strength and swell values and/or the PI (an indirect indication of the "swell potential" and "workability").

An "uncured" CBR of 12 to 15 is adequate for many construction expediting applications where the stabilized layer is to serve as a "working platform". Lower CBR values (but not less than approximately 8) may be satisfactory in some situations.

For PI reduction and workability improvement applications, design lime content is the lime percentage beyond which further increases in lime content do not effect significant changes in PI. In some instances lower lime contents may produce acceptable PI reduction and satisfactory workability. Generally the first increment of lime (< 3 percent) produces very substantial decreases in PI with increased percentages (> 3 percent) being less beneficial. Many soil-lime mixtures are non-plastic with 3 percent lime while others retain PI at increased treatment levels. It should be noted that soil modification with low percentages of lime may not provide permanent effects (as discussed in Chapter 6). Stabilization permanency requires a strength evaluation.

Proposed Mixture Design Process

Different procedures are used for structural layer applications and subgrade modification. For structural layer design, follow the flow diagram illustrated in Figure 5.1.

Subgrade modification depends on stabilization objectives. Soil-lime mixtures should be prepared at various lime percentages (2 percent increments are generally used) and tested.

Eades and Grim Procedure

The pH procedure developed by Eades and Grim (1966) is based on the philosophy of adding sufficient lime to a soil to satisfy cation exchange capacity of the soil and satisfy all initial or short term reactions and yet still provide enough lime and a high enough pH to sustain the strength-producing lime-soil pozzolanic reactions. These

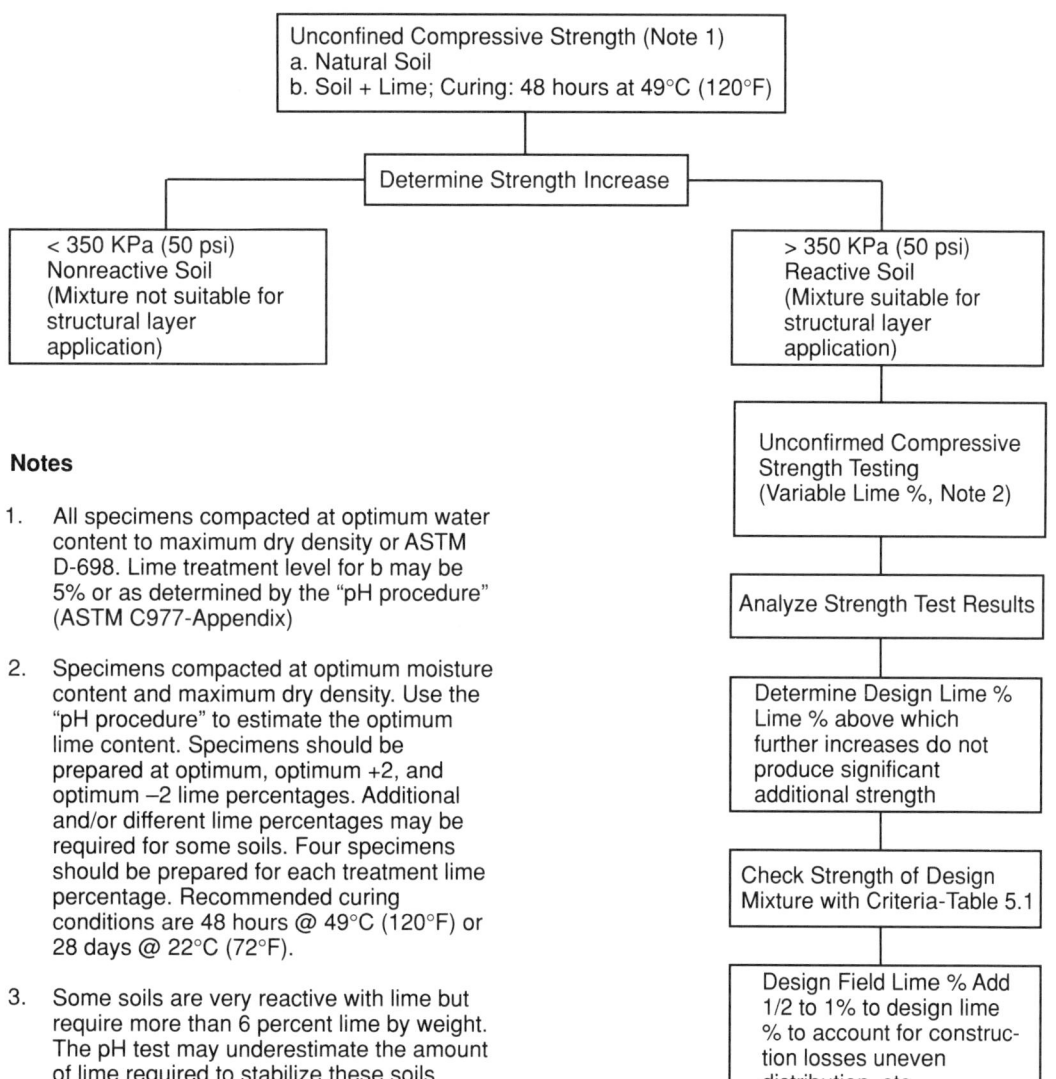

FIGURE 5.1. THE THOMPSON MIXTURE DESIGN FLOW CHART IS BASED ON SOIL-LIME REACTIVITY (AFTER LITTLE ET AL. (1987)).

reactions continue for essentially as long as the pH remains high and lime and pozzolans are available. The procedure is presented in the appendix to ASTM C-977 and is summarized below:

1. Representative samples of air-dried, minus No. 40 soil to equal 20 grams of oven-dried soil are weighed to the nearest 0.1 gram and poured into 150 ml (or larger) plastic bottles with screw tops.
2. Because most soils require between 2 and 5 percent lime, it is advisable to set up five bottles with lime percentages of 2, 3, 4, 5 and 6. This will ensure, in most cases, that the percentage of lime required can be determined in 1 hour. Weigh the lime to the nearest 0.01 gram and add it to the soil. Shake to mix the dry soil and lime.
3. Add 100 ml of CO_2-free distilled water to the bottles.
4. Shake the lime-soil and water until there is no evidence of dry material on the bottom. Shake for a minimum of 30 seconds.
5. Shake the bottle for 30 seconds every 10 minutes.
6. After 1 hour transfer part of the slurry to a plastic beaker and measure the pH. The pH meter must be equipped with a Hyalk electrode and standardized with a buffer solution with a pH of 12.00.
7. Record the pH for each of the soil-lime mixtures. If the pH readings go to 12.40, the lowest percentage of lime that gives a pH of 12.40 is the percentage required to stabilize the soil. If the pH does not go beyond 12.30 and 2 percent lime gives the same reading, the lowest percentage that gives a pH of 12.30 is that required to stabilize the soil. If the highest pH is 12.30 and only 1 percent lime gives a pH of 12.30, additional test bottles should be started with larger percentages of lime. Figure 5.2 demonstrates the results of a pH test for a Burleson, Texas, clay.

The pH test is a good test to use as a starting point for optimum lime content selection. The pH test has been shown to provide optimum lime contents that correlate well with optimum lime contents selected from strength testing for soils from Illinois (Thompson and Eades, 1970) and from other regions of the country (Haston and Wohlgemuth, 1985). However, Hardy (1970) demonstrated that the pH test is not as effective in predicting the optimum lime content for ultisols and oxisols of the tropics and subtropics when optimum lime content is defined as the lime content which provides the maximum strength.

The major limitations of the pH test are: (a) the technique does not establish whether the soil will react with lime to produce a substantial strength increase, and (b) strength data are not generated for use in evaluating mixture quality.

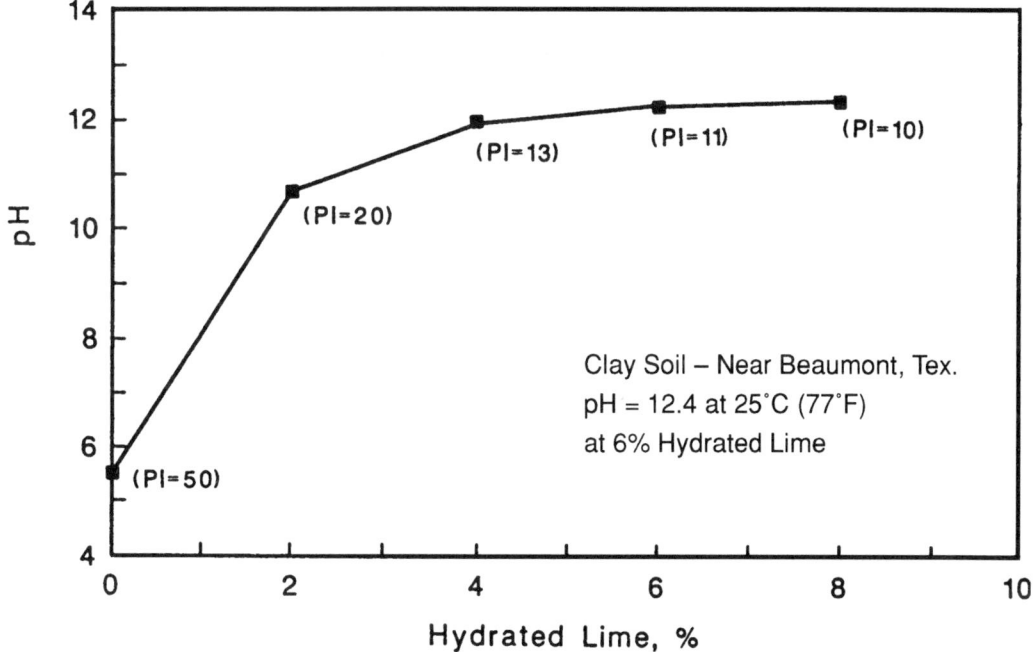

FIGURE 5.2. THE EADES AND GRIM pH TEST IS AN EXCELLENT INDICATOR OF OPTIMUM LIME CONTENT AND SHOULD BE USED AS PART OF A COMPLETE MIXTURE DESIGN PROCEDURE TO INSURE OPTIMIZATION OF POZZOLANIC REACTIVITY.

Texas Procedure

The soil-lime mixture design procedure used by the Texas Department of Transportation is AASHTO T-220, which provides for the determination of the unconfined compressive strength of the lime-soil mixture. The procedure suggests strength criteria of 690 KPa (100 psi) for base construction and 345 KPa (50 psi) for subbase construction.

Details of the procedure are presented in AASHTO T-220. A summary of the procedure is presented as follows:

1. Based on the grain size and PI data, the lime percentage is selected using Figure 5.3. The percentages in this figure should be substantiated by approved testing methods on any particular soil material. Use of the chart for materials with less than 10 percent No. 40 and cohesionless materials (PI of less than 3) is prohibited. A relatively high purity lime, usually 90 percent or more of Ca and Mg hydroxides, or both, and 85 percent or more of which passes the No. 200

sieve is required for stabilization. Percentages shown are for stabilizing subgrades and base courses where lasting effects are desired. Satisfactory temporary results are sometimes obtained by the use of as little as one-half of the aforementioned percentages.

2. Optimum moisture and maximum dry density of the mixture are determined in accordance with appropriate sections of AASHTO T-212 and Tex-113-E. The compactive effort is 50 blows of a 44.5 N (10-pound) hammer with a 45.7 cm (18-inch) drop.
3. Test specimens 15 2-cm (6-inches) in diameter and 203-cm (8-inches) in height are compacted at optimum moisture content and maximum dry density.
4. The specimens are placed in a triaxial cell (AASHTO T-212 or Tex-121-E) and cured in the following manner:
    a. Allow the specimen to cool to room temperature,
    b. Remove cells and dry at a temperature not exceeding 60°C (140°F) for about 6-hours or until one-third to one-half of the molding moisture has been removed,
    c. Cool the specimens for at least 8-hours and
    d. Subject the specimens to capillarity (AASHTO T-212 Section 6 or Tex-121-E) for 10 days.
5. The cured specimens are tested in unconfined compression in accordance with AASHTO T-212 Sections 7 and 8 or Tex-117-E.

The results of the unconfined compression strength testing can be used for substantiation of optimum lime content.

5.04 Accelerated Curing Cautions

The strength of lime-stabilized soil is both time and temperature dependent. The time required to reach a certain percentage of curing can be accelerated by curing at a higher temperature. The higher temperature accelerates the formation of pozzolanic reaction products. However, if accelerated curing temperatures are too high, the pozzolanic compounds formed during laboratory curing can differ substantially from those that would normally develop in the field. Generally, elevated temperatures in excess of 49°C (120°F) should be avoided. Recent and accumulating research evidence indicates that (40°C) 104°F at various curing times is a more appropriate temperature which accelerates curing without introducing pozzolanic products that may significantly differ from those expected during field curing.

It is common to use the compressive strength of specimens cured at 21°C (72°F) as a datum or base-line strength. However, 28 days may be an unrealistically long period of

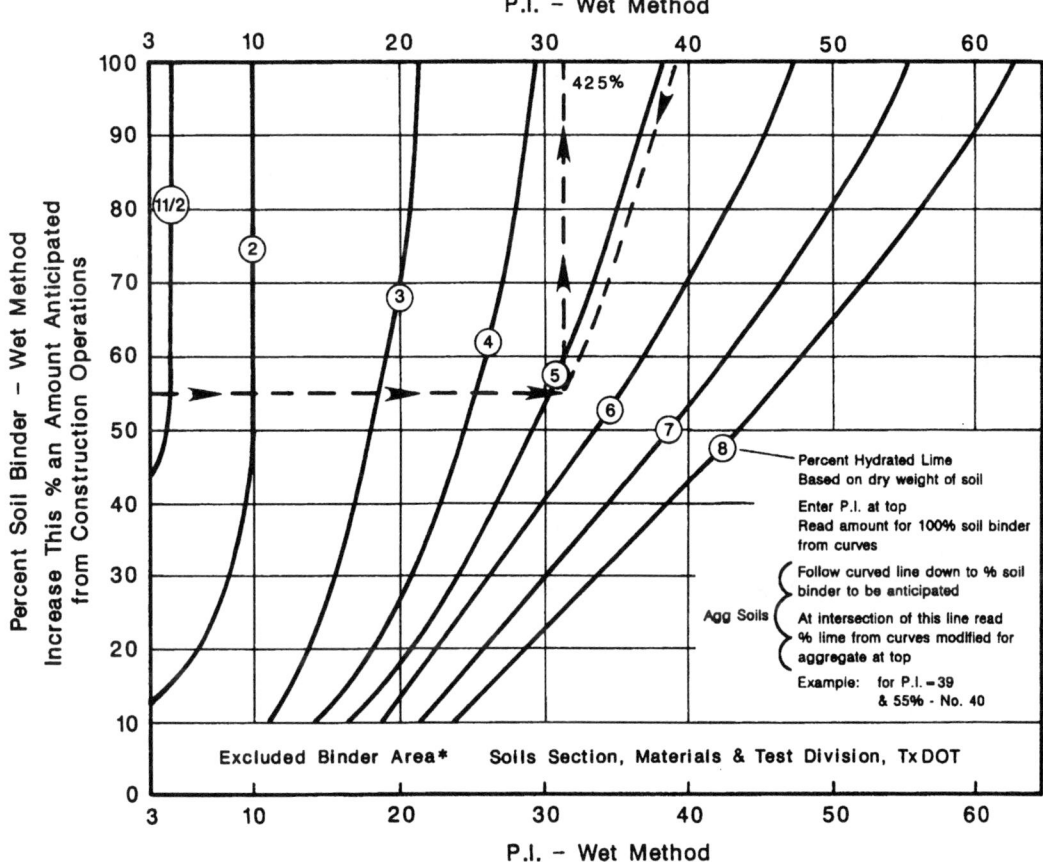

FIGURE 5.3. THE TEXAS DEPARTMENT OF TRANSPORTATION SELECTS OPTIMUM LIME CONTENT FOR STRENGTH TESTING BASED ON THE SOIL INDEX PROPERTIES OF SOIL BINDER (MINUS NO. 40 SIEVE SIZE) AND PI.

laboratory curing when mixture design results are needed quickly. Therefore, it is important to select a reasonable protocol for accelerated curing.

Although accelerated strength may be used as the basis for mixture design and quality control, it is always a good idea to verify the accelerated strength with strength data from samples cured for 28 days and at ambient temperature (22°C (72°F)).

5.05 Summary

The primary objective of a mixture design procedure is to determine an acceptable percentage of lime required to modify the soil so as to provide the desired construction performance characteristics (modification) or to determine the optimum percentage of lime required to optimize strength and durability (stabilization). It is critically important to add sufficient lime to optimize the development of pozzolanic reaction products and, therefore, pozzolanic strength if the stabilized layer is to be used as a structural pavement layer. Therefore, mixture design of a soil-lime mixture that will be used as a structural pavement layer should include a strength criterion for evaluation of optimum binder content.

Haston and Wohlgemuth (1985) tested 29 different soils from different locations in Texas. They compared the optimum binder contents determined for these soils using the criteria of (1) Atterberg limits, (2) soil-water-lime mixture pH and (3) unconfined compressive strength. They also considered the Texas method (Tex-121-E) for determining the optimum lime content. The conclusions of Haston and Wohlgemuth were:

1. Strength tests are the best indicator of the amount of lime to use for stabilization.
2. The pH test using the Eades and Grim procedure (ASTM C-977-83a) is a better predictor of peak strength than Atterberg limits tests.
3. Tex-121-E method for optimum lime content is often significantly below the strength test optimum, and this is often a surprise to practicing engineers who have, in the past, considered the Tex-121-E method to be conservative.

The work by Haston and Wohgemuth (1985) is in agreement with the extensive work performed by Currin, Allen and Little (1976).

Because of the importance of selecting an optimum lime content based on the criterion of strength, the Thompson procedure is recommended as an acceptable procedure for mixture design. Notice that this procedure incorporates the pH test for estimation of the optimum lime content followed by verification of the optimum lime content based on the criterion of strength.

The differentiation between lime modification and lime stabilization of soils cannot be overemphasized in mixture design. McCallister and Petry (1991) clearly demonstrated the importance of optimum lime content based on the lime content that produces optimum strength. Their data on over 1,700 samples subjected to severe leaching demonstrated that lime content based on optimum strength are required for "permanency" as discussed in Chapter 6. Similar findings are presented by Townsend and Klym (1965) in their suggestions for stabilizing Canadian soils to withstand freeze-induced heave. They recommend a lime content of 4 percent above the lime fixation point (the point at which PI is reduced the maximum amount for the soil in question) to insure durability in a freeze-thaw environment.

Durability and permanency of reactions are discussed in Chapter 6.

Appendix 5.01
Compression Strength of Molded Soil-Lime Cylinders
(After Little et al. 1987)

A. Purpose

Many different methods are used for determination of compressive strengths of soil-lime mixtures. These methods differ in sample preparation and sample size. The proper method of strength determination must be used with the appropriate mixture design method and evaluation procedure.

1. If a procedure for compressive strength determination is not provided, then this procedure is recommended.

B. Scope

This method covers the procedures for making and testing molded cylinders of soil-lime mixtures to determine their compressive strength. This method provides for specimens 50-mm (2-inches) in diameter and 100-mm (4-inches) in length. However, the same procedure can be adopted to different sized specimens.

C. Applicable Documents

ASTM Standards: D3551, Method for Laboratory preparation of Soil-Lime Mixtures Using a Mechanical Mixer and D2216, laboratory Determination of Moisture Content of Soil.

D. Apparatus

1. Compression Test Specimen Molds
   Molds having an inside diameter of 50-mm (2-inches) and a height of 100-mm (4-inches) for molding test specimens. The mold shall have an extension collar assembly made of rigid metal and constructed so it can be securely attached to or detached from the mold. The extension collar assembly shall have a height extending above the top of the mold of at least 50-mm (2-inches), which may include an upper section that flares out to form a funnel provided there is at least a 12-mm ($^1/_2$-inch) straight cylindrical section beneath it.
2. Compaction Hammer
   A manually operated metal hammer having a 49.3 ± 0.25 mm (1.94 + 0.01) inch diameter circular face equipped with a 17.8 N (4-pound) rammer that slides freely on a metal rod attached to the circular compaction face. The rammer shall have a drop of 305-mm (12-inches).

3. Compression Specimen Extruder
   A device consisting of a piston, jack and frame or similar equipment suitable for extruding specimens from the mold.
4. Scarifying Tool
   A sharp-edged or sharp pointed device suitable for scarifying the surface of a compacted soil-lime layer.
5. Miscellaneous Equipment
   Tools such as spatulas, trowels, scoops, etc. for use in preparing the specimens.
6. Compressive Device
   The compression device may be any device with sufficient capacity and control to provide a constant strain rate which may range from 0.50 to 2.0 percent per minute. The device shall be equipped in such a manner that the compressive load can be applied to the specimen without producing eccentric loading conditions. The compression device shall be capable of measuring the unit load to the nearest 13.8 KPa (2 psi).

E. *Preparation of Soil-Lime Mixture*

1. The mixture shall be prepared in accordance with ASTM D3551.

F. *Molding Specimens*

1. Three specimens shall be prepared.
2. Compact the mixture into the mold in three approximately equal layers using the compaction hammer. The surfaces of the first two layers should be scarified to promote bonding between adjacent layers. The compaction effort (number of blows per layer) is selected to provide the desired density.

Note: A compacted effort of 20 blows per layer produces densities approximately equal to ASTM D698.

3. Trim the compacted soil-lime mixture even with the top of the mold by means of a straightedge.
4. Extrude the specimen from the mold, determine the mass of the specimen and record the mass.
5. Take a moisture content sample from the remaining soil-lime mixture after the second specimen has been compacted.
6. Cure the specimen in the manner desired. Constant temperature curing for a designated time period is normally used. Typically, curing is carried out in sealed containers to avoid moisture loss and lime carbonation.

G. *Compression Test*

   1. Place the specimen in the compression device making certain that the specimen is properly aligned.
   2. Apply the load continuously and without shock so as to produce axial strain at a rate of 0.5 to 2.0 percent per minute. Record the maximum load sustained by the specimens to the nearest 26.7 N (6 pounds).
   3. Determine the moisture content of a representative sample from the three specimens tested.

H. *Calculation*

   1. Calculate the compressive strength by dividing the maximum load by the cross-sectional area of the specimen.
   2. Determine the average compressive strength of the three specimens tested.

I. *Report*

   1. The report shall include the following:
      a. Mixture identification (percent lime, soil sample identification, lime identification).
      b. Length of mellowing period used in mixture preparation in accordance with ASTM D3551.
      c. Specimen diameter and length (in millimeters) and cross-sectional area (in square millimeters).
      d. Strain rate used (percent per minute).
      e. Average compressive strength (calculated to the nearest 13.8 KPa (2 psi)).
      f. Curing conditions (time, hours; temperature, degrees Celcius; nature of curing container).
      g. Moisture content (in percent) and dry density (in g/cm^3) at molding.
      h. Moisture content (in percent) of the specimens after the test.

5.06 References

Currin, D. D., Allen, J. J. and Little, D. N., (1976). "Validation of Soil Stabilization Index System with Manual Development, Report No. FJSRL-TR-76-0006, Frank J. Seiler Research Laboratory, United States Air Force Academy, Colorado.

Eades, J. L., and Grim, R. E., (1966). "A Quick Test to Determine Lime Requirements for Soil Stabilization," *Highway Research Record* No. 139.

Hardy, J. R., (1970). "Factors Influencing Lime Reactivity of Tropically and Subtropically Weathered Soils," Ph.D. Dissertation, University of Illinois—Champaign—Urbana.

Haston, J. and Wohlgemuth, S.K., (1985). "Comparisons of Compressive Strengths of Lime-Soil Mixtures Based on Optimum Lime Contents Determined By Different Methods."

Little, D. N., Thompson, M. R., Terrel, R. L., Epps, J. A. and Barenberg, E. J., (1987). "Soil Stabilization for Roadways and Airfields," Report No. ESL-TR-86-19, Engineering and Services Center, Tyndall Air Force Base, Florida.

McCallister, L.D., and Petry, T. M., (1991). "Property Changes in Lime Treated Expansive Clays Under Continuous Leaching," Technical Report GL-90-17.

Transportation Research Board, (1987). *State of the Art Report No. 5, Lime Stabilization.*

Thompson, M. R., (1970). "Suggested Method of Mixture Design Procedure for Lime-Treated Soils," American Society for Testing Materials, Special Technical Publication 479.

Thompson, M. R. and Eades, J. L., (1970). "Evaluation of Quick Test for Lime Stabilization," *Journal of the Soil Mechanics and Foundations Division,* American Society of Civil Engineers.

Townsend, D. L. and Klym, T. W., (1965). "Durability of Lime-Stabilized Canadian Soils," C. E. Report No. 54, Queens University, Kingston, Ontario.

CHAPTER 6

ENGINEERING PROPERTIES OF LIME-STABILIZED SOILS AND AGGREGATES

6.01 Properties and Characteristics of Soil-Lime Mixtures

In general all lime treated fine-grained soils exhibit a reduction in plasticity, decreased shrink-swell potential and improved workability (TRB State of the Art Report No. 5, 1987). However, not all soils demonstrate a substantial level of improved strength gain. This strength gain is due to pozzolanic reactivity. The level of improvement in physical properties exhibited in soils is dependent upon soil type, lime type, lime percentage and curing conditions, i.e., time, temperature and moisture.

Lime treated soils may be broken down into two categories: reactive and modified. Thompson (1966) has defined the reactive soils as those which demonstrate an unconfined compressive strength gain of at least 345 kPa (50 psi) over the untreated soil strength. The term reactive in this definition refers to pozzolanic reactivity as defined in Chapter 5. The term modified refers to soils with limited strength gain (less than 345 kPa (50 psi)) but with significant physical property changes: plasticity, volume change potential, texture and workability.

The concept of modified lime-soil mixtures versus stabilized lime-soil mixtures can be somewhat confusing when one considers the mixture design objectives of modification versus stabilization as discussed in Chapter 5. In mixture design terminology, any soil, reactive or non-reactive, can be modified by adding the appropriate lime content to alter selected physical properties (Chapter 5), or to modify the soil by reducing plasticity, shrink-swell potential or by improving workability and constructability. In the context of mixture design most soils (reactive soils) can also be stabilized by adding enough lime to promote long-term strength gain. Thompson (1966) points out that certain soils can only be modified by lime addition because they are not pozzolonically reactive.

Although soils that do not demonstrate significant strength gain are not usually considered for use as a structural pavement layer, it should be understood that the reduction in plasticity, textural change and reduction in volume change potential occurring in these soils results in improved stiffness or resilient moduli in these soils. These modified soils should provide better and less variable roadbed support as discussed in Chapter 3.

Lime treatment results in both immediate and long-term effects (TRB State of the Art Report No. 5, 1987) on soil properties. This discussion of engineering properties is divided into two categories: immediate or "uncured" mixture properties and long-term or "cured" mixture properties. The curing period refers to a period of time when temperature and moisture are sufficient to provide an adequate environment for pozzolanic strength gain.

Uncured Mixtures

Plasticity

Substantial reduction in plasticity is caused by lime treatment, and the soil often becomes non-plastic. Generally, high PI and high clay content soils require greater quantities of lime for achieving the nonplastic condition, if it can be achieved. The first increments of lime added are most effective in reducing plasticity (TRB State of the Art Report No. 5, 1987). The silty and friable texture of the treated soil causes a marked increase in workability. The improved level of workability expedites subsequent manipulation and placement of the treated soil.

Table 4.1 (page 41) summarizes immediately identifiable changes in liquid limit and PI for soils from Illinois and Texas as a function of the percentage of lime added. Figure 6.1 demonstrates the change in plasticity index for a Texas soil, a California soil and two South Dakota soils as a function of the percentage of lime added to the soil.

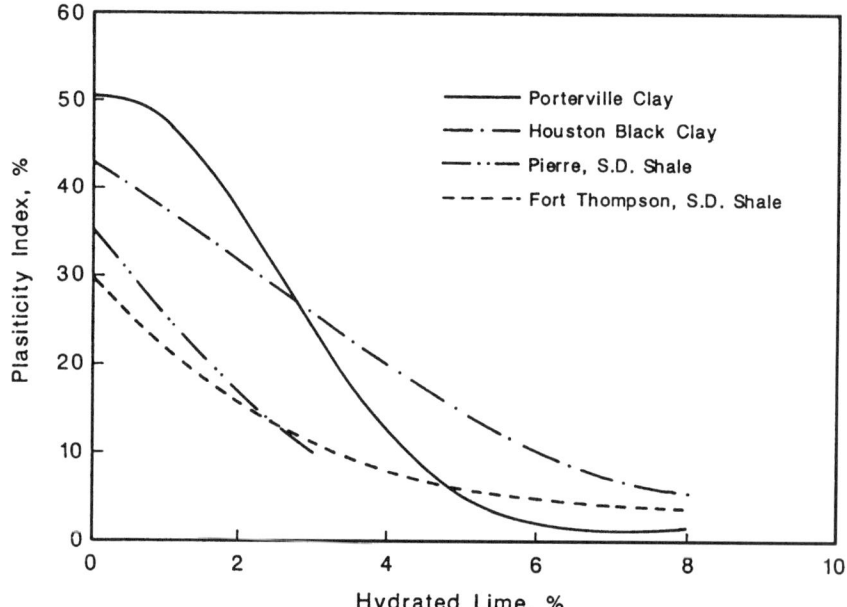

FIGURE 6.1. DIFFERENT PERCENTAGES OF LIME ARE REQUIRED TO REDUCE PLASTICITY TO DESIRED LEVELS FOR DIFFERENT SOILS. THIS REACTION IS IMMEDIATE IN THAT IT DOES NOT REQUIRE LONG CURING TIMES. (AFTER HOLTZ, 1969).

The soils in Figure 6.1 are pozzolanically reactive. Figure 6.2 demonstrates the changes in Atterberg limits for four additional Texas soils including a Beaumont Clay which develops pozzolanic strength gain but only with extended (long-term) curing. Notice that although this soil does not develop a significant pozzolanic strength gain immediately, it does immediately demonstrate substantial and valuable physical property changes which influence the behavior and performance of the soil in the pavement structure.

Moisture Density Relationships

The result of immediate reactions (i.e., cation exchange, flocculation/agglomeration and early calcium-aluminate-hydrate formation caused by Ca and $Ca(OH)_2$ crowding at the surface of the clay mineral) between lime and the soil is a substantial change in the moisture density relationship. Figure 6.3 illustrates this point for a CL soil (Little et al, 1987).

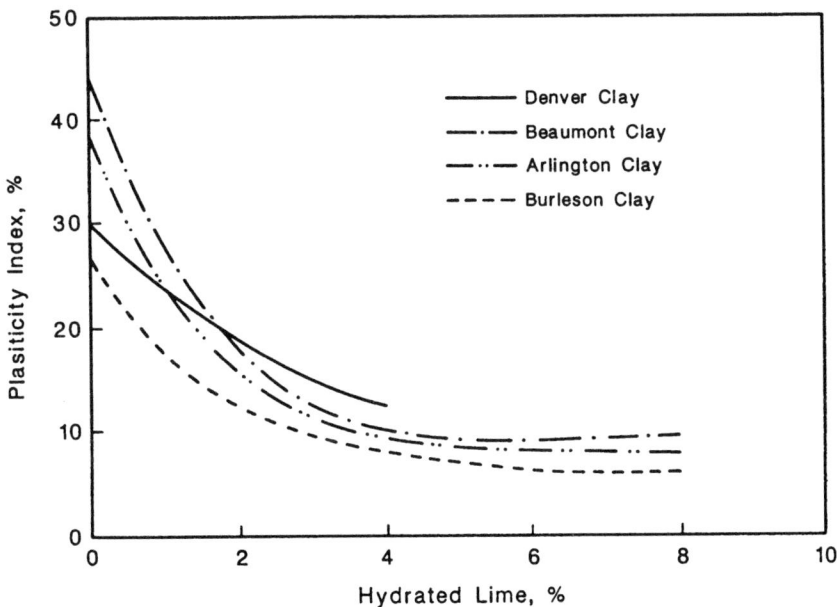

FIGURE 6.2 EVEN SOILS WHICH REQUIRE LONG-TERM CURING (AT LEAST 28-DAYS) TO DEVELOP SIGNIFICANT STRENGTH GAINS DEMONSTRATE IMMEDIATE PI REDUCTION AS DEMONSTRATED BY THIS BEAUMONT CLAY.

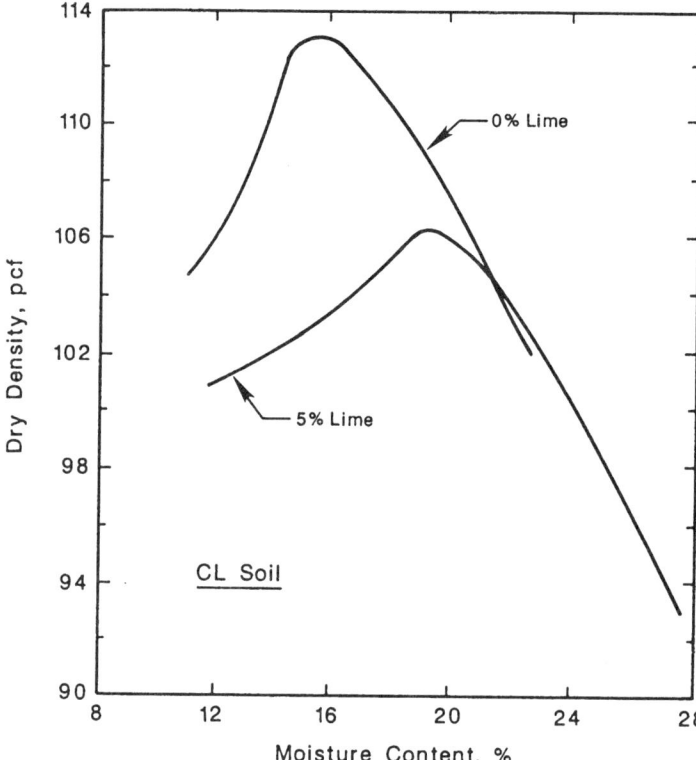

FIGURE 6.3 THE SHIFT IN DENSITY AND OPTIMUM MOISTURE CONTENT FOR ACHIEVING MAXIMUM DENSITY IS EVIDENCE OF THE PHYSICAL CHANGES THAT OCCUR (IMMEDIATELY) DURING LIME TREATMENT. (AFTER TERREL ET AL., (1979)).
(1 pct = 16Kg/m^3)

These moisture-density changes reflect the new nature of the soil and are evidence of the physical property changes occurring in the soil upon lime treatment. For a specific compactive effort, lime-treated soil has a lower density and a higher optimum moisture content than does the untreated soil. The reduction in maximum dry density is typically from 48 to 80 Kg/m^3 (3-5 pounds per cubic foot) with a typical increase in optimum moisture content of 2-4 percent, Little et al. (1987). However, in highly plastic clays, substantially greater increases in optimum moisture may be realized.

If a mixture is allowed to cure and gain strength prior to compaction, further reduction in maximum dry density and an additional optimum moisture content increase may be noted. It is important that the appropriate moisture-density curve, in terms of percent lime used and time of curing, be used for field control purposes.

Swell Potential

Soil swell potential and swelling pressure are normally significantly reduced by lime treatment. In fact, the reduction in PI associated with virtually all fine-grained soils upon the addition of lime is a significant indication of the reduction of swell potential due to lime stabilization. A relationship between PI and swell potential was developed by Seed, Woodward and Lundgren (1962) which states:

Percent Swell = $0.00216 \times PI^{2.44}$ (6.1)

This relationship is also plotted in Figure 6.4. In this plot the percent swell is defined as volume change incurred in the soil as the moisture content increases from optimum moisture content to saturation moisture content.

Table 6.1 presents California Bearing Ratio (CBR) and CBR swell data for several Illinois soils and some Texas soils. The swell is determined during the 96 hour soak period which is part of the CBR test procedure. Although the swell potential of lime-treated soils vary, it is common for lime treatment to reduce the swell to less than 0.1 percent.

Associated with a reduction in volume increase or swell is decrease in swell pressure. Figure 6.5 illustrates how swell pressure of a lime-treated clay is reduced as the percentage of hydrated lime added to the clay is increased. This clay is a CH material with a PI of 36. This figure illustrates both the influence of the amount of lime used and the

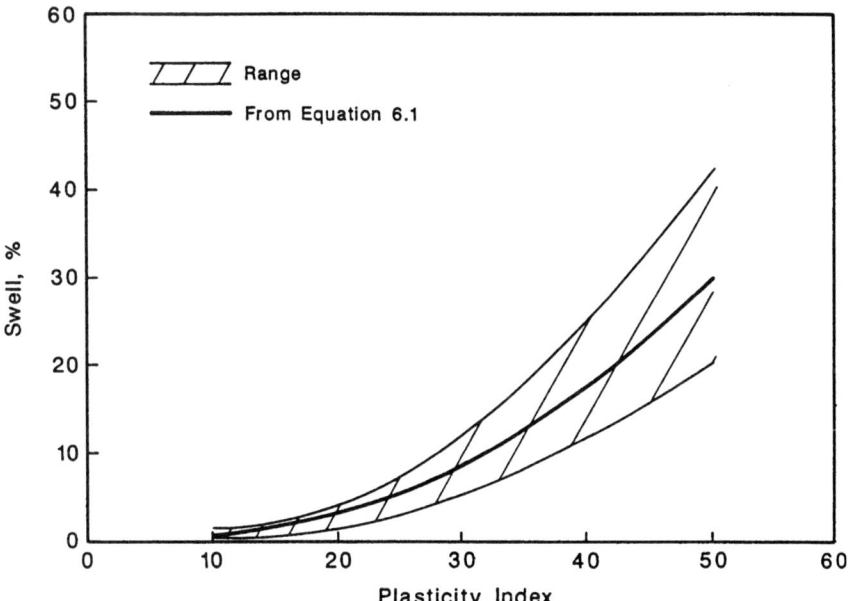

FIGURE 6.4. PLASTICITY INDEX CAN BE USED TO PREDICT SWELLING AS DEMONSTRATED BY SEED ET AL. (1962).

Table 6.1. CBR Values for Natural and Lime-Treated Soils (After Thompson, 1969).

Soil	Unified Classification	Natural Soil		% Lime	Soil-Lime Mixtures			
					No Curing		48 Hours Curing @120°F	
		CBR, %	Swell, %		CBR, %	Swell, %	CBR, %	Swell, %
Reactive Soils								
Accretion Gley 2	CL	2.6	2.1	5	15.1	0.1	351.0	0.0
Accretion Gley 3	CL	3.1	1.4	5	88.1	0.0	370.0	0.1
Bryce B	CH	1.4	5.6	3	20.3	0.2	197.0	0.0
Champaign Co. Till	CL-ML	6.8	0.2	3	10.4	0.5	85.0	0.1
Cisne B	CH	2.1	0.1	5	14.5	0.1	150.0	0.1
Cowden B	CH	7.2	1.4	3	—	—	98.5	0.0
Cowden B	CH	4.0	2.9	5	13.9	0.1	116.0	0.1
Cowden C	CL	4.5	0.8	3	27.4	0.0	243.0	0.0
Darwin B	CH	1.1	8.8	5	7.7	1.9	13.6	0.1
East St. Louis Clay	CH	1.3	7.4	5	5.6	2.0	17.3	0.1
Fayette C	CL	1.3	0.0	5	32.4	0.0	295.0	0.1
Illinoian B	CL	1.5	1.8	3	29.0	0.0	274.0	0.0
Illinoian Till	CL	11.8	0.3	3	24.2	0.1	193.0	0.0
Illinoian Till	CL	5.9	0.3	3	18.0	0.9	213.0	0.1
Sable B	CH	1.8	4.2	3	15.9	0.2	127.0	0.0
Non-Reactive Soils								
Fayette B	CL	4.3	1.1	3	10.5	0.0	39.0	0.0
Miami B	CL	2.9	0.8	3	12.7	0.0	14.5	0.0
Tama B	CH	2.6	2.0	3	4.5	0.2	9.9	0.1

[1] Specimens were placed in 96 hours soak immediately after compaction.

influence of curing time. Goldberg and Klein (1952) illustrated the influence of compaction density on swell pressure. They demonstrated that an increase in compaction density from 1,440 Kg/m^3 to 1,632 Kg/m^3 (90 to 102 pcf) increased swell pressure of a Porterville, California, clay by 90 percent. However, a similar increase in compaction density had a negligible effect on the same soil when stabilized with 8 percent Ca(OH)$_2$.

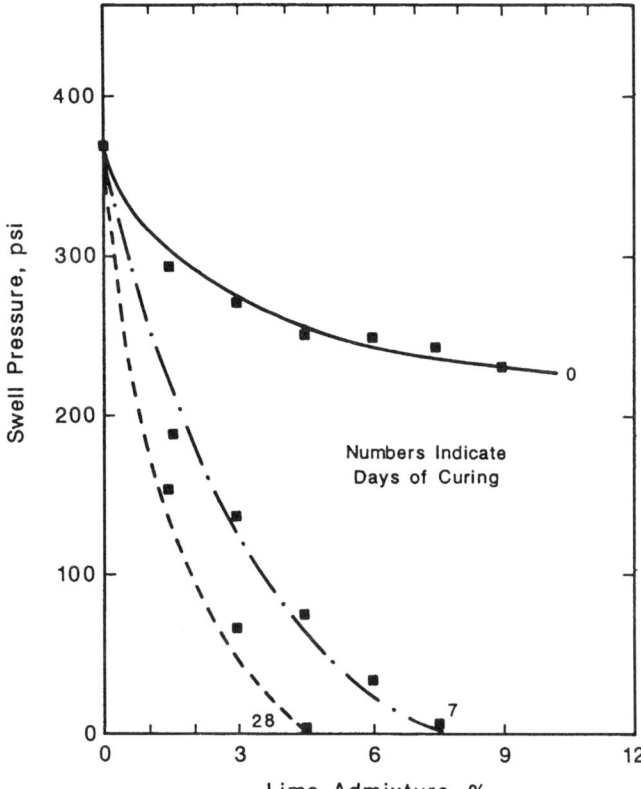

FIGURE 6.5. SWELL PRESSURE AS A FUNCTION OF LIME CONTENT AND PERIOD OF CURING FOR IRBID, JORDAN, CLAY (AFTER BASMA AND TUNCER, 1991).

Textural Changes

A further illustration (Figure 6.6) of the effect of lime on producing physical property changes is reflected in the reduction in the clay-size fraction of the same soil discussed in Figure 6.5. Note the dramatic reduction in percent smaller than 2 micrometer material. This is primarily due to the effect of flocculation and agglomeration.

Strength and Deformation Properties

Lime-treatment of fine-grained soils produces immediate improvement in strength and deformation properties of "uncured" soil-lime mixtures. These immediate benefits are evident from CBR, cone index, R-value, static compression and resilient modulus testing.

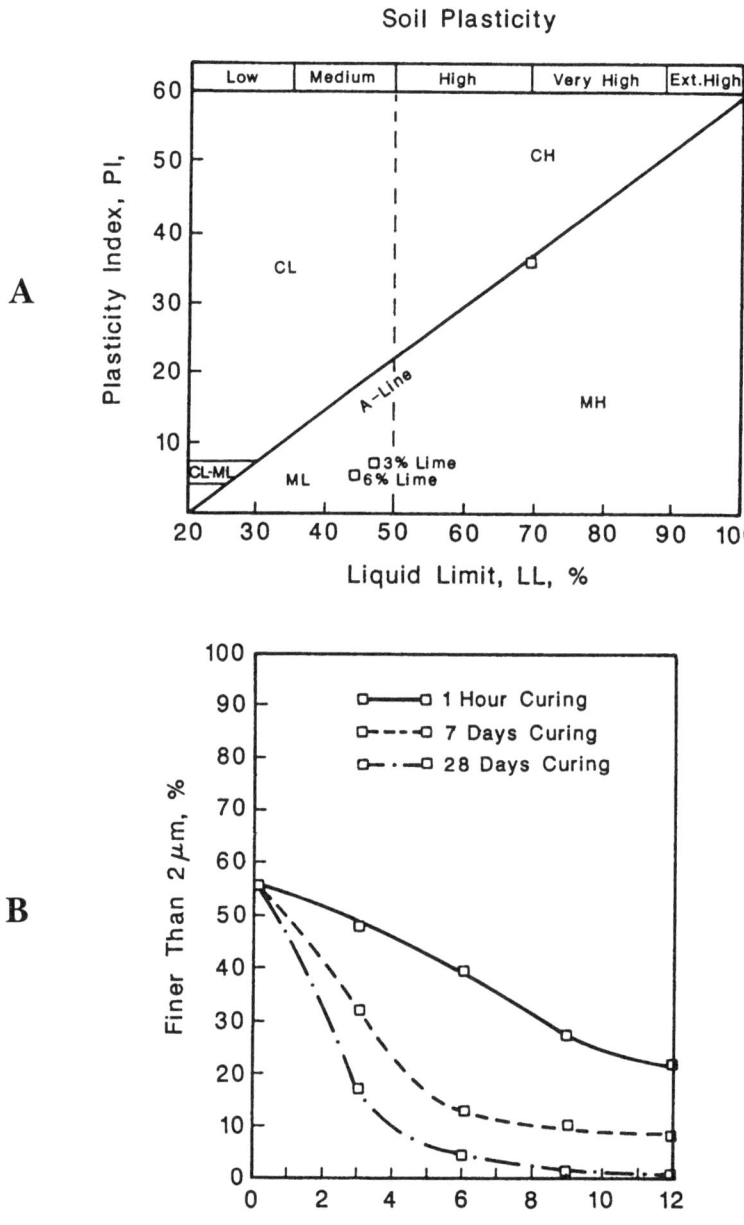

FIGURE 6.6 EFFECT OF LIME AND CURING TIME ON THE (A) PLASTICITY AND (B) CLAY-SIZE FRACTION OF IRBID, JORDAN CLAY.
(AFTER BASMA AND TUNCER, 1991).

Typical moisture-content versus CBR relations of uncured soil-lime mixtures and the natural soil are shown in Figure 4.7, page 42. The arrows on this figure indicate the optimum moisture content of each mixture. Note that as the percent lime is increased, the optimum moisture content is increased. Using optimum moisture content as a reference, note that the immediate CBR of the low plasticity clay illustrated in Figure 4.7 is increased from 8.5 to 13.5 percent with 3 percent lime and to 14.5 percent with 5 percent lime. Furthermore, it is important to note that at the higher moisture content of 20 percent as a reference, the strength as measured by CBR increases from approximately 3 percent to approximately 12 percent, a 400 percent increase.

Figure 6.7 illustrates the resilient modulus of a natural soil, a low plasticity silt, and the same soil with 3 percent lime. The resilient modulus is plotted versus repeated deviator stress to illustrate the stress sensitivity of fine-grained soils. Of greatest significance in this figure is the fact that the modulus is increased from approximately 17,235 kPa (2,500 psi) to approximately 172,350 Pa (25,000 psi) at a deviatoric stress level of 34 kPa (5 psi), which is typical of the deviatoric stress induced in a roadbed (subgrade) soil. Since this increase in modulus is the result of "uncured" conditions, it is illustrative of the significant changes in roadbed deformation properties which can occur due to lime-treatment even without significant long-term, pozzolanic reactions. The significance of roadbed modulus is discussed further in Chapters 3 and 7.

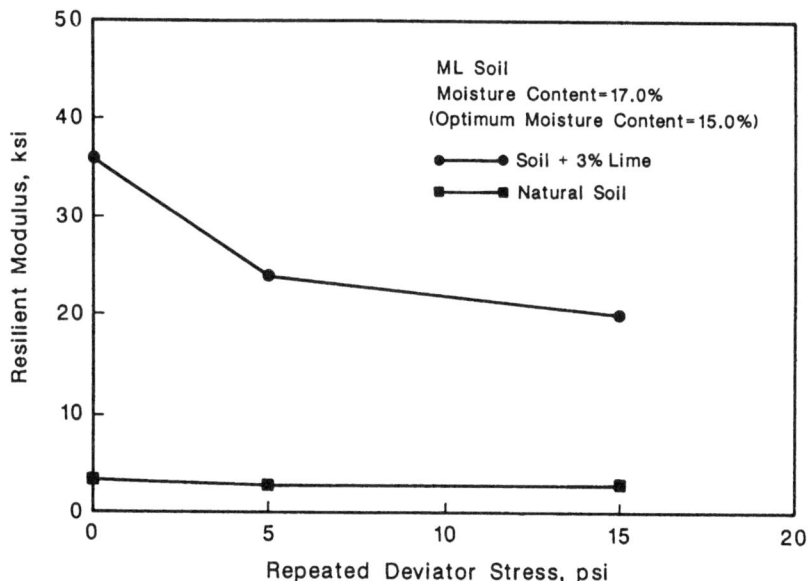

FIGURE 6.7. IMMEDIATE EFFECTS OF LIME TREATMENT CAN BE SUBSTANTIAL AND ARE DEMONSTRATED BY THIS 10-FOLD IMPROVEMENT IN RESILIENT MODULUS. (AFTER LITTLE ET AL. (1987)).
NOTE: DATA COLLECTED BY M. R. THOMPSON, UNIVERSITY OF ILLINOIS, 1982.

Summary

The immediate effects of lime treatment on soils which are suitable for lime stabilization (fine-grained soils with at least 10 percent clay and a plasticity index of 10 or more) are due to the mechanisms of cation exchange, $Ca(OH)_2$ adsorption to the clay surface and to some extent rapid development of pozzolanic products. The level of physical property changes in the soil which results from lime treatment is quite soil dependent. However, virtually all fine-grained soils, regardless of soil-lime pozzolanic reactivity, derive some level of physical property consistency improvement through lime treatment as reflected by changes in Atterberg limits and changes in volumetric measurements due to moisture fluctuations.

Cured Mixtures

Strength Properties

The most important effect of long-term curing is the development of pozzolanic products as discussed in Chapter 4. The development of more pozzolanic products results in more "glue" to hold the particles of soil together and a mineralogical change favorable to greater strength.

Figure 6.8 illustrates strength data versus lime content for five Texas soils and a Denver, Colorado, soil. Table 6.2 represents similar data for the same five soils shown in Figure 6.8 as well as for 20 Illinois soils. These data coupled with the unconfined compressive strength data versus lime content in Figure 4.9 (page 47) for several different clay mineral compositions indicate the importance of selection of the appropriate mixture design and lime content to maximize long-term strength properties. Data from Figure 6.8 also illustrate the long-term strength gain effects of lime. Note especially that the Beaumont clay does not demonstrate reactivity during accelerated curing (7 days at 38°C (100°F)) but does demonstrate considerable strength gain during longer periods of ambient cure.

Lime stabilized soils tend to gain strength at a slower rate than portland cement-stabilized soils. This slower strength gain should be considered in mixture design and pavement design. Longer term curing data, 28 to 360 days, are presented in Table 6.3 for 12 California soils. The California soils were cured at 23°C (73°F). A very interesting point with these California soils is that most of the soils are relatively low plasticity, clayey silts or silty clays (Doty and Alexander, 1978). Yet they are very reactive as is evidenced by the substantial strength gain. This indicates that even low percentages of a reactive clay may result in considerable strength gain.

Tuncer and Basma (1991) used a ratio of the lime treated soil strength to the strength of the natural soil (LSR) as a measure of the effect of lime percent and curing time on strength gain. They found that the LSR of a Jordanian CH clay stabilized with 9 percent lime increased from 3 at 4 days to 7 at 28-days curing.

Table 6.2. Compressive Strength Data for Natural and Lime-Treated Soils from Illinois, Texas and Colorado (Modified After Little et al. (1987)).

Soil	Unified Classification	Compressive Strength, psi Percent Lime		
		3	5	7
Arlington, TX	CH	250	350	650
Beaumont, TX	CH	70	100	200
Burleson, TX	CH	150	220	310
Victoria, TX	CH	100	190	260
Denver, CO	CL	300	400	350
Bryce A, IL	MH	43	58	53
Bryce B, IL	CH	201	212	193
Cisne B, IL	CH	107	190	189
Drummer A, IL	ML	29	49	32
Drummer B, IL	CH	186	152	146
Fayette A, IL	ML	37	46	49
Fayette B, IL	CL	109	114	113
Fayette C, IL	CL	137	185	125
Accretion-Gley, IL	CL	263	247	283
Huey B, IL	CL	223	216	233
Huey D, IL	CL	222	179	197
Illinoian Till, IL	CL	150	186	143
Loam Till, IL	MH	172	184	174
Davidson B, IL	MH	198	268	324
Greenville B, IL	CL	455	517	551
Norfolk B, IL	SC	347	421	332
Clalitos B, IL	MH	114	133	132
Nipe B, IL	ML	87	220	311
Cecil B, IL	CH	168	163	224
St. Ann Bauxite, IL	CH	104	292	495

Note: Curing Conditions of 28 days at 22°C (73°F).
Data for all Illinois soils provided by M. R. Thompson (1982).
1 psi = 6,894 Pa

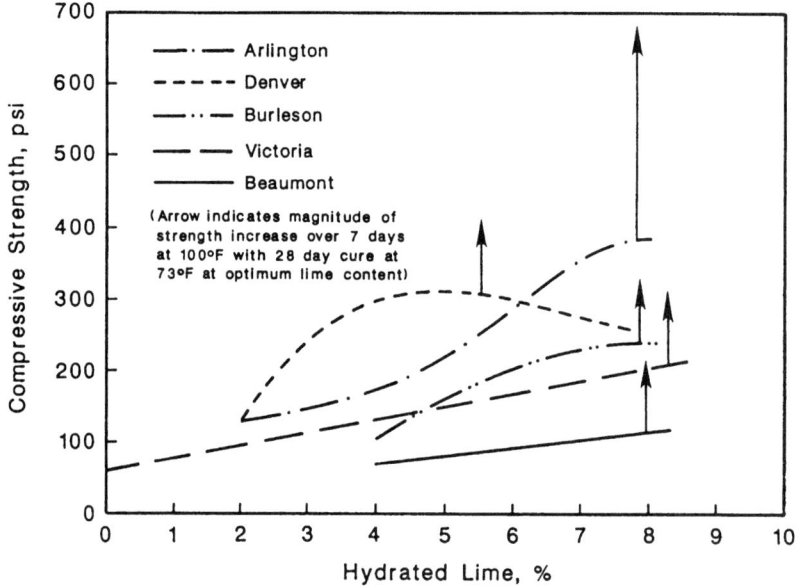

FIGURE 6.8. ALTHOUGH ACCELERATED CURING PROCEDURES CAN BE USED TO APPROXIMATE LONG-TERM STRENGTH GAIN, LIME-SOIL POZZOLANIC REACTIONS OCCUR OVER TIME. THE LONG-TERM BENEFICIAL POZZOLANIC EFFECTS SHOULD BE ACCOUNTED FOR IN DESIGN.

The very significant strength increases illustrated in Table 6.3 and in Figure 6.8 further illustrate the importance of maintaining enough lime and a high enough pH to continue pozzolanic reactivity in a reactive soil. As long as this is maintained, continued strength gain is possible. In fact, field data indicate that, under proper conditions, strength gain can continue for in excess of 10 years.

Although the unconfined compressive strength is the most widely used measure of the shear strength of lab-fabricated lime-stabilized soils, other measures are also used. Among these are the CBR, California R-value and triaxial shear strength.

CBR

CBR values for many cured lime-soil mixtures are high and indicate the extensive development of pozzolanic cementing products. For mixtures with CBR's of 100 or more, test results have little practical significance (TRB State of the Art Report No. 5, 1987). In this case the unconfined compressive strength is preferred as an indicator of

Table 6.3. Change in Unconfined Compressive Strength During Curing for Twelve California Soils (After Doty and Alexander, 1978).

			Unconfined Compressive Strength, psi					
	AASHTO	Plasticity	3 Percent Lime			7 Percent Lime		
Soil	Classification	Index	28 Day	180 Day	360 Day	28 Day	180 Day	360 Day
1	A-6 (10)	14	160	210	220	120	210	610
2	A-6 (10)	11	390	410	510	400	120	1410
3	A-7-5 (20)	30	280	360	310	550	1190	1210
4	A-2-4	NP	100	100	100	110	150	180
5	A-7-6 (20)	30	350	450	640	260	1200	1650
6	A-7-5 (13)	15	70	60	70	220	200	220
7	A-4 (5)	7	80	160	280	120	210	400
8	A-6	14	540	700	750	550	1200	1580
9	A-4	7	420	920	1100	350	1250	1900
10	A-7-5 (20)	22	400	760	830	300	950	1200
11	A-4 (2)	10	275	410	900	210	800	1110
12	A-7-5 (20)	22	360	430	520	510	810	1010

1 psi = 6,894 Pa

reactivity and of strength. This is because the unconfined compressive strength is a measure of compressive shear strength (a basic material property) and is presented in engineering units which can be used in engineering calculations, analysis and design. In addition, other important engineering properties can be directly derived from the unconfined compressive strength such as tensile strength and flexural strength. Both of these properties are important in pavement design and analysis.

Despite the limitations of CBR testing, it is an indirect measure of shear strength and can be used to rank order or prioritize strength gain.

Table 6.1 is a valuable illustration of the changes in CBR as an indicator of soil shear strength in uncured and cured soils.

Triaxial Strength

In some cured soil-lime mixtures the major effect of lime on the shear strength is to produce a substantial increase in cohesion with some minor increase in friction angle (φ angle). Typical angles of shearing resistance (φ) are approximately 25° to 35° (TRB State of the Art Report No. 5, 1987). Cohesion increases with increased mixture compression strength. A rough estimate of cohesion is approximately 30 percent of the unconfined compressive strength (TRB State of the Art Report No. 5, 1987).

Tuncer and Basma (1991) have documented the improvement in undrained shear strength properties of a typical plastic Jordanian clay. Their data illustrate the significant increase in undrained cohesion and in undrained angle of internal friction, ϕ_u, as a function of the amount of lime added and time of curing, Figures 6.9a and 6.9b.

A modified version of a triaxial test is used by the Texas Department of Transportation to evaluate the quality of aggregate and soil-aggregate materials for use as subbase and base courses. This triaxial apparatus is illustrated in Figure 6.10. Varying levels of lateral confining pressure are applied to a sample and the axial compressive stress is increased for each confining pressure until shear failure occurs. The result of the series of tests defines a failure envelope. When this failure envelope is plotted on the design-acceptance chart, Figure 6.11, the mixture is classified as to its acceptability depending on the relative position of the failure envelope with respect to the evaluation chart.

A Colorado River bank run gravel (from near Columbus, Texas) comprised of a crushed, siliceous river gravel and a high PI (approximately 35) clay binder was stabilized with 2.5 percent hydrated lime. After a curing period of only 7-days, the triaxial classification was a Class 2 material. Treatment with 5 percent lime improved the triaxial classification to Class 1. The untreated soil was classified as a Class 4 material (see Figure 6.11). The 96-hour soak laboratory CBR of the lime stabilized Colorado River Gravel with 5 percent hydrated lime was 100 percent as compared to a CBR of 30 for the untreated river gravel.

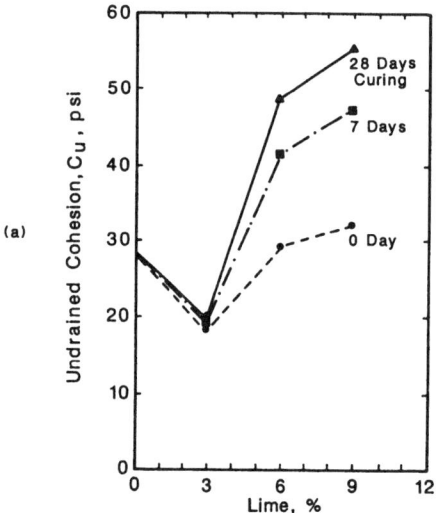
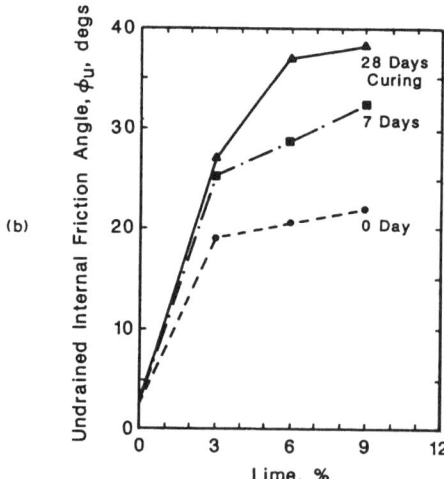

FIGURE 6.9. EFFECT OF LIME CONTENT AND CURING TIME ON UNDRAINED COHESION (A) AND UNDRAINED ANGLE OF INTERNAL FRICTION (B) FOR JORDANIAN SOIL (AFTER TUNCER AND BASMA, 1991).
1psu = 6,894 Pa

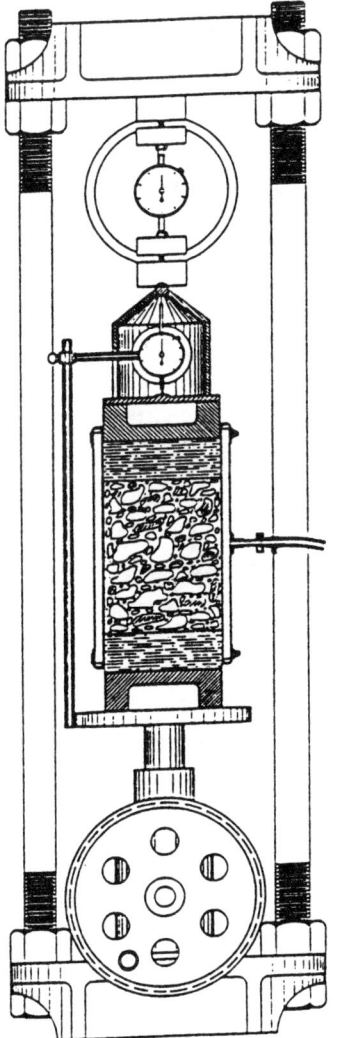

FIGURE 6.10. TEXAS DEPARTMENT OF TRANSPORTATION USES A TRIAXIAL TEST TO EVALUATE THE ACCEPTABILITY OF GRANULAR (UNBOUND AND STABILIZED) BASES AND SUBBASES.

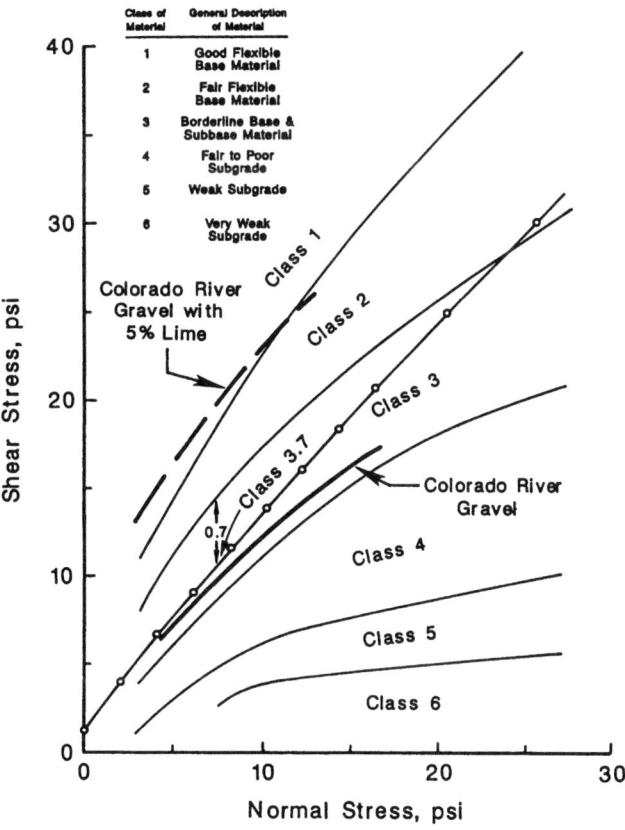

FIGURE 6.11. LIME STABILIZATION OF A COLORADO RIVER GRAVEL, RICH IN CLAY BINDER, IMPROVES THE TEXAS TRIAXIAL CLASSIFICATION.

1 psi = 6,894 Pa

Tensile Strength

As the unconfined compressive strength of a lime-soil mixture increases the tensile strength does also. Two test procedures are commonly used to measure the tensile strength of lime-soil mixtures: the indirect tensile or splitting tensile test and the flexural beam test. Large variations are common in indirect tensile testing, and these variations are dependent on the nature of the lime-soil mixture and the curing conditions.

The ratio of tensile strength to unconfined compressive strength is approximately 0.13, and this is a strong enough correlation to be used for normal design purposes (Little et al., 1987).

The most common method used for evaluating the flexural tensile strength of highway materials is the flexural test (beam strength) as this value can be related to the stabilized slab which bends under the action of traffic loading in the field. A realistic estimate of the flexural strength (modulus of rupture) is 0.25 times the cured unconfined compressive strength of the mixture (Little et al., 1987).

Fatigue Strength

The fatigue strength and fatigue life of lime-soil mixtures are linked to the critical flexural stress induced within the stabilized layer and to the flexural tensile strength or modulus of rupture of the mixture. Cured soil-lime mixture flexural fatigue response curves are comparable to those normally obtained for similar materials (with regard to the nature of the cementitious products) such as lime-fly ash and aggregate mixtures and soil cement mixtures and concrete. In order to withstand a large number (i.e., approximately 5 million or more) of cyclic loadings resulting in flexural stresses in the stabilized layer, it is necessary that the flexural strength of the stabilized mixture and layer be approximately twice the flexural stress induced per loading cycle.

Thus an approach to thickness design of a lime-stabilized structural layer to reduce fatigue potential would be to design the layer or the entire pavement to be thick enough, stiff enough or otherwise structurally able to reduce the critical flexural stress within the stabilized layer so that it is less than about 50 percent of the rupture modulus of the lime-stabilized layer. For a lime-soil mixture with an unconfined compressive strength of 2,758 kPa (400 psi), the flexural strength can be approximated as 25 percent of the unconfined compressive strength or 688 kPa (100 psi). Thus the pavement should be designed so that the flexural stress at the bottom of the lime-stabilized layer is no more than 344 kPa (50 psi) in order to insure a fatigue life of approximately 5 million applications of the design vehicle (critical wheel or axle load in the traffic mix).

Deformation or Modulus Properties

Determination of proper stress-strain properties are essential for analyzing the behavior of a pavement structure containing a lime-soil layer. Figure 6.12 (Goose Lake

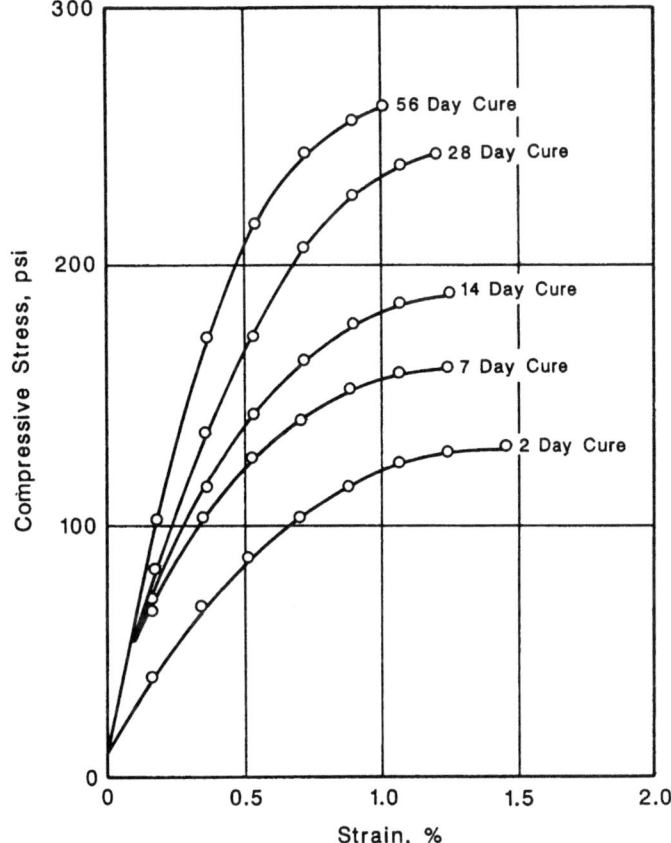

FIGURE 6.12. AS THE STRENGTH OF A SOIL-LIME MIXTURE INCREASES WITH CURING, THE STIFFNESS OF THE MIXTURE DOES ALSO (AFTER SUDDATH AND THOMPSON, 1975).

1 psi = 6,894 Pa

clay) illustrates that these stress-strain characteristics change with time for reactive stabilized mixtures such as the Goose Lake clay (Suddath and Thompson, 1975). As the lime-soil mixture continues to cure, the strength increases, and the strain at failure is reduced.

Soil-lime mixtures tested in compression are strain sensitive and the ultimate strain (for maximum compressive strain) is approximately one percent (Thompson, 1966), regardless of the soil type or curing period. This is illustrated more vividly in Figure 6.13 for a Brazos Valley, Texas, clay. Note that upon stabilization with 5 percent lime and curing for 28-days at 22°C (73°F), the unconfined compressive strength is increased from approximately 344 kPa (50 psi) to approximately 2,481 kPa (360 psi), and the strain at failure is reduced from approximately 5 percent to approximately one percent.

The compressive static modulus of elasticity can be estimated from the unconfined compressive strength of the lime-soil mixture according to the following relation (Thompson, 1966):

E(ksi) = 10 + 0.124 (Unconfined Compressive Strength in psi)

Resilient moduli (as that measured by AASHTO T-274) are typically 2 to 3 times higher than the static modulus.

It should be considered that the deformation properties, moduli, like all other strength properties of lime-soil mixtures are dynamic and thus change with time and further curing. As curing continues and if pozzolanic activity continues, the flexural strength and stiffness or modulus increases. The result is improved strength and fatigue life properties with increased strength gain during extended curing.

Poisson's Ratio

Poisson's ratio, like modulus, is a stress dependent property. At low stress levels (less than 50 percent of the ultimate compressive strength), it is generally in the range of 0.1 to 0.2. At higher stress levels, Poisson's ratio may be closer to the 0.2 to 0.3 range. A value of between 0.15 and 0.20 is typically used (TRB State of the Art Report No. 5, 1987).

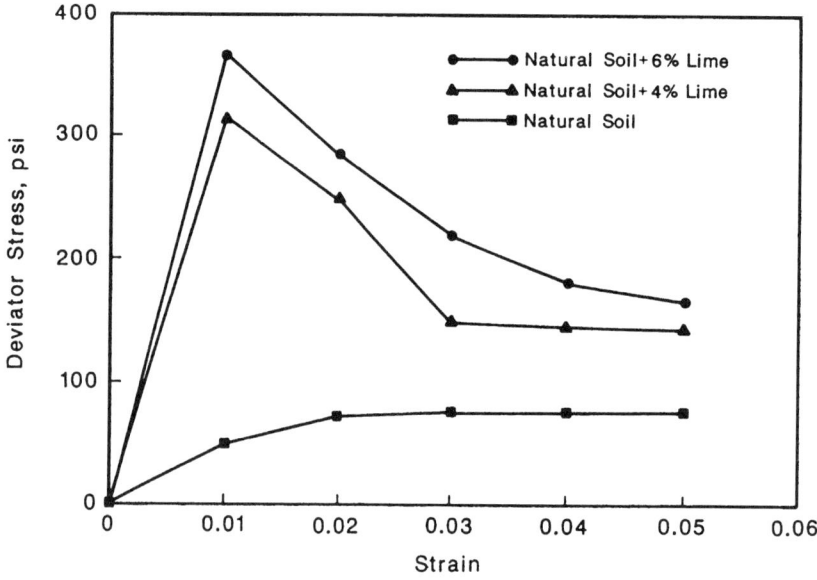

FIGURE 6.13. THE STRESS-STRAIN CURVE OF THIS BURLESON, TEXAS, CLAY IS SUBSTANTIALLY CHANGED BY LIME STABILIZATION. THE SLOPE OF THE INITIAL PORTION OF THE CURVE REPRESENTS THE STIFFNESS OR MODULUS OF THE MIXTURE (AFTER SHRINIVAS, 1992).

Permeability

The permeability of soil can change significantly in response to three effects: mineral dissolution, ion exchange reactions and desiccation due to displacement of water by highly organic fluids. The addition of lime to a soil and water system produces the first two effects as explained in Chapter 4. The results of the effects of mineral dissolution and ion exchange, when the ion is Ca^{++}, is an increase in permeability, at least initially. In fact Townsend and Klyn (1966) found significant permeability increases upon lime treatment of soils and related this increase in permeability to the increase in pore volume due to flocculation. Ranganatham (1961) found a 10-fold increase in permeability in lime-treated expansive clays. Other researchers (McCallister and Petry, 1990) have found that initially lime-treated soils demonstrate an increase in permeability followed by a decrease because of pozzolanic product which accumulates in the interstitial regions. The permeability of many of these lime-stabilized soils remain considerably higher than that of the natural soil. However, in some lime-reactive soils, the permeability decreases with curing and associated pozzolanic reactions to approximately that of the natural soil.

McCallister and Petry (1990) found that the permeabilities of three expansive North Central Texas soils were from 7 to 300 times higher after lime treatment than for the natural clays without lime stabilization. They further found that the permeability decreased with leaching, especially if the lime content used to stabilize the soils was a low percentage of lime (3 to 4 percent). However, if the lime content was optimum for strength gain (6 to 7 percent), the permeability changes upon leaching were negligible.

The effect of leaching on permeability is probably related to the loss of calcium ions during the leaching process. As the calcium (electrolyte) concentration is reduced, the particle repulsion is increased and water retention at the clay surface is increased. The result can be a higher level of retained moisture around the clay particles and lower permeability. It is important to note that the long-term permeability is affected by the prevalent cation in the soil water system and the concentration of that cation. It is also important to note that the development of pozzolanic reaction products resist permeability alternations upon leaching (McCallister and Petry, 1990).

Shrinkage of Lime-Soil Mixtures

Shrinkage of stabilized soils due to the loss of moisture can result in shrinkage cracking within the stabilized soil which can further result in reflection cracking in asphalt concrete surfaces resting over the stabilized layer. Lime treatment of clay soils reduces the shrinkage potential of these soils as the lime-clay reaction results in mineralogical modification of the clay to provide a more moisture stable structure as discussed in Chapter 4.

Field moisture content data for lime-treated soils suggests that the moisture content changes in the stabilized material are not large when the water content and stabilizer content used in construction is approximately the optimum content of each (Little et al., 1987).

Calculations based on laboratory shrinkage data, as well as field service data from many areas, indicate that, for typical field service conditions, shrinkage of cured soil-lime mixtures will not be extensive. Thus, reflective cracking through the surface course should not frequently occur if proper construction techniques are used (Little et al., 1987).

Durability of Lime-Soil Mixtures

The primary durability considerations concerning lime-soil mixtures deals with prolonged exposure to moisture and cyclic freeze-thaw effects. Extensive work by Thompson and Dempsey (1969) on Illinois soils reveals that the ratio of soaked to unsoaked compressive strengths for lime-soil mixtures is typically between 0.7 and 0.85. Lime-soil mixtures seldom reach 100 percent saturation. The maximum degree of saturation is typically in the range of 90 to 95 percent.

Pavement systems may experience two general types of freeze-thaw action. Cyclic freeze-thaw occurs in the material when freezing occurs as the advancing frost line moves by and then thawing subsequently occurs. Heaving conditions develop when a quasi-equilibrium frost-line condition is established in the stabilized material layer. The static frost line situation provides favorable conditions for moisture migration and subsequent ice lens formation and heaving, if the material is frost-susceptible. Depending on the nature of the prevailing climate in the area, either cyclic freeze-thaw or heaving action or both may occur.

In zones where freezing temperatures occur, freeze-thaw damage may be incurred by the soil-lime mixtures. The damage is generally characterized by volume increase and strength reduction.

Initial unconfined compressive strength (exposed to no freeze-thaw cycles) of a cured lime-soil mixture is a good indicator of freeze-thaw resistance. Durability studies of several different types of "cemented" materials (including lime-soil, portland cement-soil and lime-fly ash-soil) have confirmed that initial compressive strength of the cured mixture can be used to predict the cyclic freeze-thaw resistance of stabilized soils. Factors influencing strength development (curing time, density, additive content, etc.) influence cyclic freeze-thaw resistance in the same fashion.

The cured soil-lime mixture must be sufficiently strong prior to the initiation of cyclic freeze-thaw action to withstand the freeze-thaw strength loss. Freeze-thaw durability considerations must, therefore, be considered in establishing mixture compressive strength requirements.

Soil-lime mixtures display autogenous healing properties. If the stabilized soil has the ability to regain strength or "heal" with time, the distress produced during winter freeze-thaw cycles will not be cumulative, since autogenous healing during favorable curing conditions would restore the stability of the material. The phenomenon of autogenous healing is demonstrated in Figure 6.14 (Thompson and Dempsey, 1969) where strength gain occurs to substantial levels in periods of curing following periods of freeze-thaw damage accumulation. In order for this "healing" to occur in the periods, not only must temperature be adequate but the components necessary to continue the pozzolanic reaction must be in place: i.e., residual calcium from an adequate supply of lime and a high enough system pH to release clay silica and clay alumina.

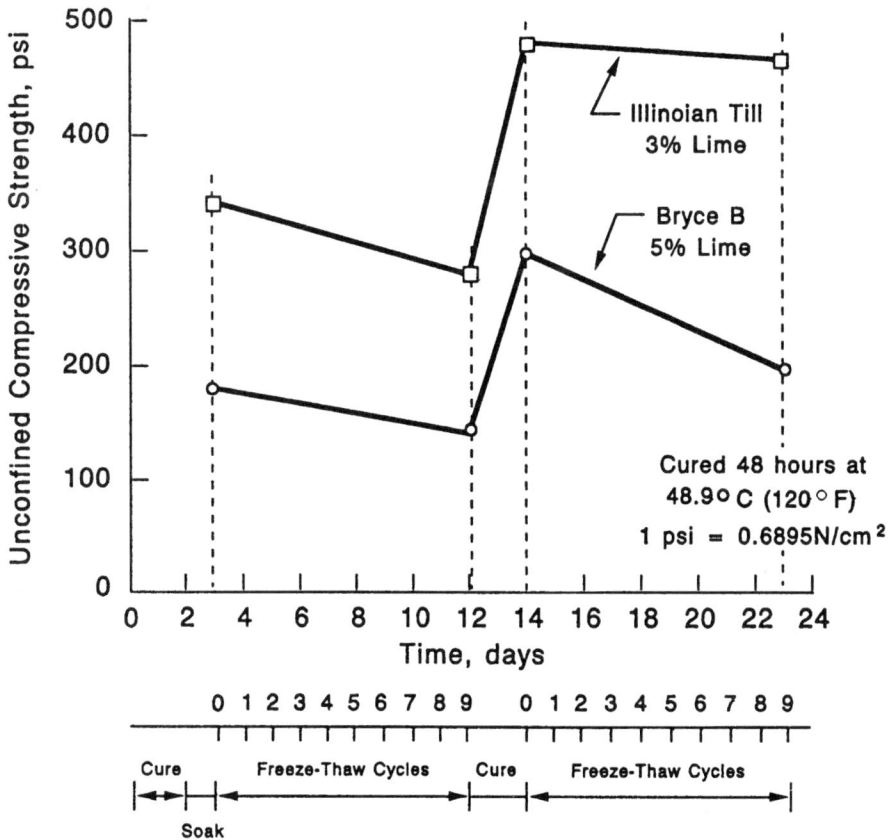

FIGURE 6.14. IF ADEQUATE LIME IS AVAILABLE, POZZOLANIC REACTION WILL CONTINUE TO OCCUR UNDER FAVORABLE CURING CONDITIONS RESULTING IN AUTOGENOUS HEALING DURING FAVORABLE SEASONS (AFTER THOMPSON AND DEMPSEY, 1969).

If "cemented" systems achieve a certain critical mixture strength level, the tensile strength of the stabilized mixture is great enough to withstand the heaving pressures generated, and heaving is eliminated or limited to tolerable values. Thompson (1970) suggests that lime-soil mixtures with compressive strengths greater than 1,379 kPa (200 psi) generally display adequate heave resistance. Townsend and Klym (1962) determined minimum strength requirements for Canadian soils to resist expansive forces resulting from freeze-thaw effects. Their study recommended a minimum compressive strength of 1,379 kPa (200 psi) (tensile strength of approximately 172 kPa (25 psi)) for active clayey (CH) soils and a minimum unconfined strength of 2,068 kPa (300 psi) (tensile strength of 310 kPa (45 psi)) for inactive silty clay and silt soils (CL). They further recommended minimum curing periods of 2 to 4 weeks for active clayey soils and 2 to 4 months for inactive silty soils. Finally, Townsend and Klym (1962) recommended stabilization with a lime content of at least 4 percent in excess of the lime fixation percentage. This lime fixation percentage is defined as that percentage of lime which causes the soil's plastic limit to reach a stable value, i.e., no appreciable changes with addition of lime. With these Canadian soils, this fixation usually was at about 2 to 3 percent lime. Thus the recommended design lime content was approximately 6 to 7 percent lime.

Figure 6.14 illustrates the influence of freeze-thaw cycles on unconfined compressive strength and the relationship between initial compressive strength and residual strength after various numbers of freeze-thaw cycles.

Thompson (1970) has demonstrated that average rates of strength decrease for a wide range of mixtures were 62 kPa (9 psi) per freeze-thaw cycle following 48 hours of curing at 49°C (120°F) and 124 kPa (18 psi) per cycle following 96 hours of curing at 49°C (120°F).

Compression

The rate of compression, or consolidation, is usually governed by the rate at which pore water can escape from the soil. One parameter commonly used to define the rate of consolidation is Cv, which is defined as:

$$C_v = \frac{K}{M_v V_w}$$

where K is coefficient of permeability, Mv is coefficient of volume change, and Vw is unit weight of water. Alternately,

$$C_r = \frac{T_r^2 d}{T\alpha c}$$

where T_r is time factor, d is one-half the thickness of the specimen for two-way drainage and $t_{\alpha c}$ is time to α-percent consolidation (α is taken to be 50 percent ($T_r = 0.196$). Then with d being essentially constant, Cr is simply a function of $t_{\alpha c}$.

Figure 6.15 defines the effect of lime treatment on the time to half full consolidation, for a Jordanian soil (PI-36).

Variability of Lime-Soil Mixtures

Liu and Thompson (1966) have tested replicate samples of lime-soil mixtures of various types in unconfined compression testing, indirect tensile testing and flexural testing. The coefficient of variation of the properties for the various mixtures was in the 11 to 12 percent range. This level of variability is typical of other paving materials: asphalt concrete, portland cement stabilized soils and portland cement concrete.

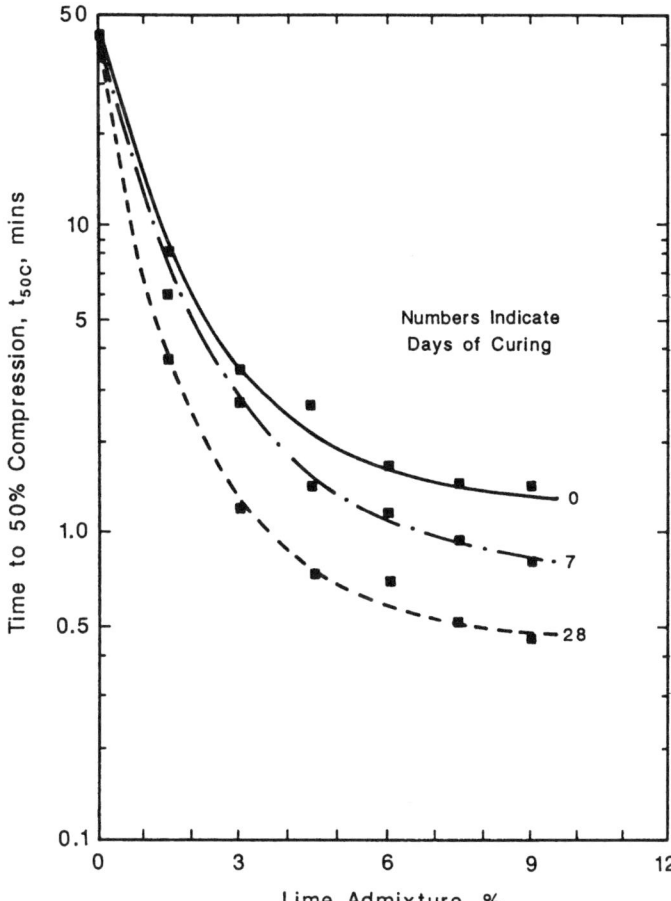

FIGURE 6.15. EFFECT OF LIME AND CURING TIME ON RATE OF COMPRESSION FOR JORDANIAN SOIL (AFTER BASMA AND TUNCER, 1991).

6.02 Engineering Properties of Lime-Soil Mixtures Under Field Conditions (in situ) and Under Simulated Field Loading Conditions (lab)

General

The ratio of applied stress to induced strain (or unit deformation) is a critically important parameter in pavement design and analysis. This ratio can be determined from a variety of tests including compressive tests, flexural tests, static tests and dynamic tests. Of these one of the most appropriate in terms of pavement design is a dynamic test in which the load is applied in a uniaxial compressive mode. This test is called a resilient modulus test. It is appropriate when testing or characterizing soil, aggregates and stabilized soil and aggregate systems. This test is defined in AASHTO test procedure T-274.

The values of resilient moduli of soils and aggregate systems when tested under the format explained in AASHTO T-274 vary widely for soils and aggregates, and the resilient modulus is highly sensitive to the stress state, the molding moisture content and conditioning moisture content of the soil. Figures 6.16 and 6.17 illustrate typical resilient moduli relationships for coarse-grained soils and aggregates, such as those used in bases and subbases; and fine-grained, cohesive soils, typical of many subgrades, respectively.

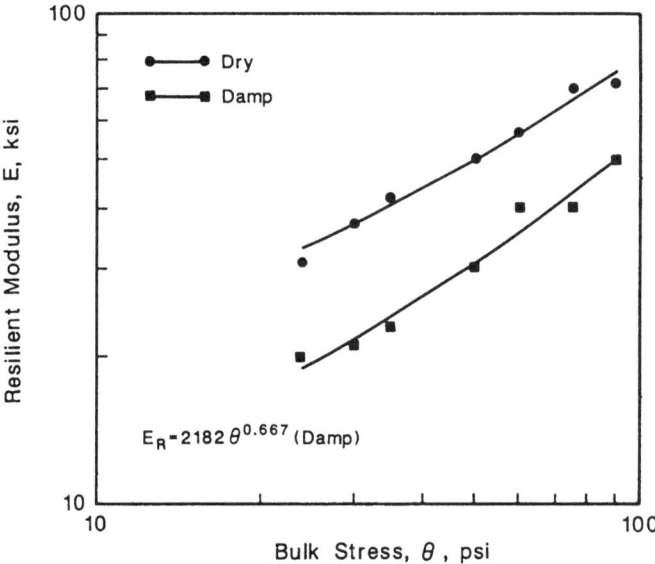

FIGURE 6.16. THE RESILIENT MODULUS OF GRANULAR SOILS AND AGGREGATES IS A FUNCTION OF THE STRESS STATE AND MOISTURE CONTENT AND IS TYPICALLY PRESENTED AS A FUNCTION OF THE BULK STRESS.

1 psi = 6,894 Pa

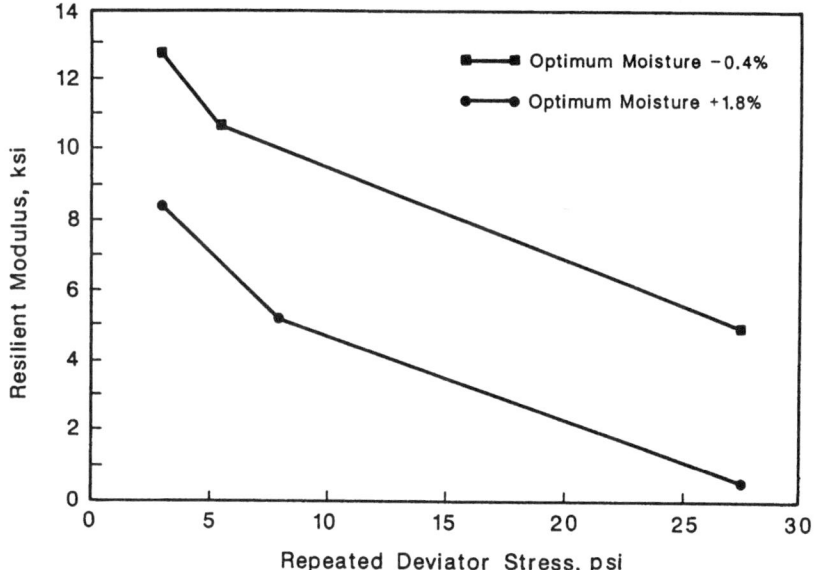

FIGURE 6.17. THE RESILIENT MODULUS OF FINE-GRAINED SOILS IS A FUNCTION OF THE STRESS STATE AND MOISTURE CONTENT AND IS TYPICALLY PRESENTED AS A FUNCTION OF DEVIATOR STRESS.
1 psi = 6,894 Pa

According to the 1986 AASHTO Pavement Design Guide, the range of resilient moduli of good quality aggregate bases is from approximately 55,152 kPa (8,000 psi) to approximately 555,152 kPa (80,000 psi). The magnitude of the resilient modulus depends on the state of stress developed within the aggregate base and on the moisture content and state of drainage within the aggregate base. The range in effective average annual resilient moduli for most subgrade soils is from approximately 34,470 kPa (5,000 psi), for cohesive subgrade soils, to as high as 103,410 kPa to 137,880 kPa (15,000 to 20,000 psi) for sandy and gravelly soils. Of course, the fine-grained, cohesive soils vary greatly in terms of resilient modulus on an annual basis as depicted in Figure 6.18. This variation in modulus is due to a variation in the seasonal moisture content and other environmental effects. In situ moduli of fine-grained subgrade soils subjected to freezing in the winter and rapid thaw in the spring may fluctuate as much as 5,000 percent during approximately a one-month period.

As a point of reference Table 6.4 presents effective resilient moduli of roadbed soils deemed to be very poor, poor, fair, good and very good for six regions of the U. S. These regions are shown in Figure 6.19.

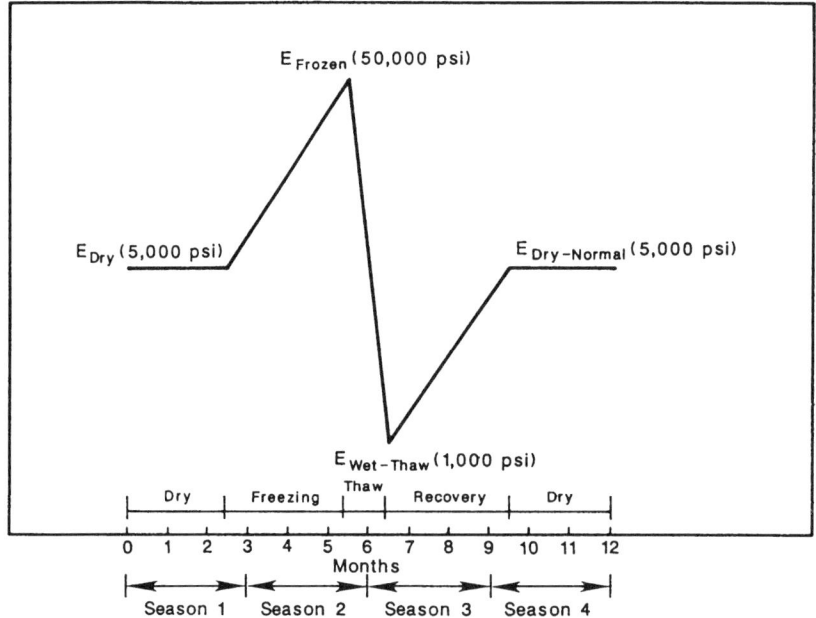

FIGURE 6.18. ILLUSTRATION OF SEASONAL VARIABILITY OF RESILIENT MODULUS OF A FINE-GRAINED, COHESIVE SOIL.
1 psi = 6,894 Pa

Table 6.4. Effective Roadbed Soil Resilient Modulus Values, M_r (psi), That May Be Used in the Design of Flexible Pavements for Low-Volume Roads. Suggested Values Depend on the U.S. Climatic Region and the Relative Quality of the Roadbed Soil (After AASHTO Pavement Design Guide, 1986).

U.S. Climatic Region	Relative Quality of Roadbed Soil				
	Very Poor	Poor	Fair	Good	Very Good
I	2,800*	3,700	5,000	6,800	9,500
II	2,700	3,400	4,500	5,500	7,300
III	2,700	3,000	4,000	4,400	5,700
IV	3,200	4,100	5,600	7,900	11,700
V	3,100	3,700	5,000	6,000	8,200
VI	2,800	3,100	4,100	4,500	5,700

* Effective Resilient Modulus in psi
1 psi = 6,854 Pa

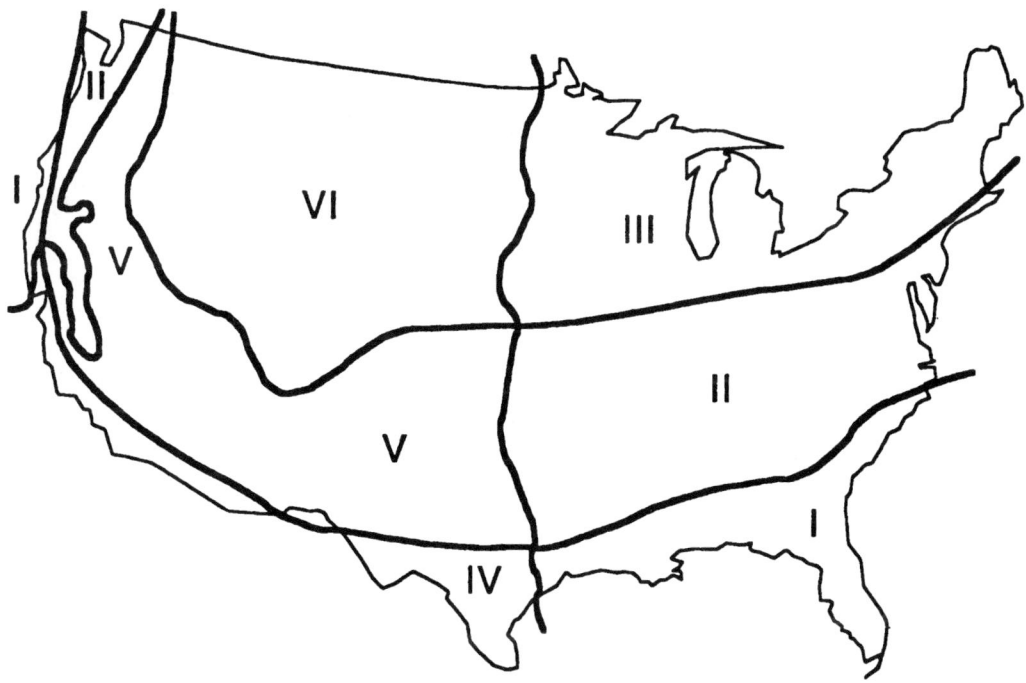

Region	Characteristics
I	Wet, no freeze
II	Wet, freeze-thaw cycling
III	Wet, hard-freeze, spring thaw
IV	Dry, no freeze
V	Dry, freeze-thaw cycling
VI	Dry, hard freeze, spring thaw

FIGURE 6.19. AASHTO HAS DIVIDED THE UNITED STATES INTO SIX CLIMATIC REGIONS FOR PAVEMENT DESIGN AND ANALYSIS (AFTER AASHTO PAVEMENT DESIGN GUIDE, 1986).

The effect of lime stabilization is to add cohesive strength to the soils and aggregate systems through plasticity reduction and pozzolanic development. This results in a substantial stiffening effect which is reflected in the in situ resilient modulus. Fossberg (1969) evaluated resilient moduli for cohesive montmorillonitic soils prepared at relatively high moisture contents and low levels of densification. Despite the fact that these soils were prepared at relatively high moisture contents and low densification levels, resilient moduli of over 689 MPa (100,000 psi) were measured in many situations. This is an increase of approximately 20 times what would typically be expected for the unstabilized soils under similar conditions of moisture and densification.

Maxwell and Joseph (1967) used field vibration testing to evaluate in situ moduli of lime-stabilized subgrades and subbases. The computed moduli of the subgrade soils ranged from 1,138 MPa (165,000 psi) immediately following construction to over 3,792 MPa (550,000 psi) two years after construction. The moduli of the subbase material ranged from 1,351 MPa (196,000 psi) immediately following construction to over 6,894 MPa (1,000,000 psi) two years after construction.

In Situ Moduli of Aggregate Base Course Stabilized With Low Percentages of Lime

Little (1990) used non-destructive deflection data from in situ testing to determine in place resilient moduli of various soils and aggregate systems, both unstabilized and stabilized. A Falling Weight Deflectometer (FWD) was used in the calculations. The FWD produces a force impulse on the pavement which closely simulates a moving wheel load. A changeable mass is allowed to fall from several heights along a shaft. The force of the falling weight is applied to the surface of the pavement through a set of rubber springs, onto a circular plate. The surface deflections are recorded at several positions by electronically integrating the signal from velocity transducers.

The deflection basin defined by the FWD was used by Little (1990) to back-calculate resilient moduli of several pavement sections. Figure 6.20 summarizes the average deflection basins for six pavement sections near Phoenix, Arizona. Each of the six pavements evaluated had the same pavement cross-section (i.e., same layer thicknesses and same materials used in each layer). The aggregate base course underlying the hot mix asphalt concrete pavement surface was from the same source and met the same specifications in each pavement section. However, the base course was unstabilized in pavement sections 7, 8 and 9 and was stabilized with one percent lime in sections 1, 3 and 6. The aggregate base course consisted of approximately 10 to 13 percent minus 200 sieve-sized material with a plasticity index of approximately 12 to 15 percent. The deflections portrayed in Figure 6.20 represent the average of at least 70 deflection readings for each pavement section. The resilient moduli calculated from these average deflections for the aggregate base course (ABC) for each pavement section are summarized in Table 6.5.

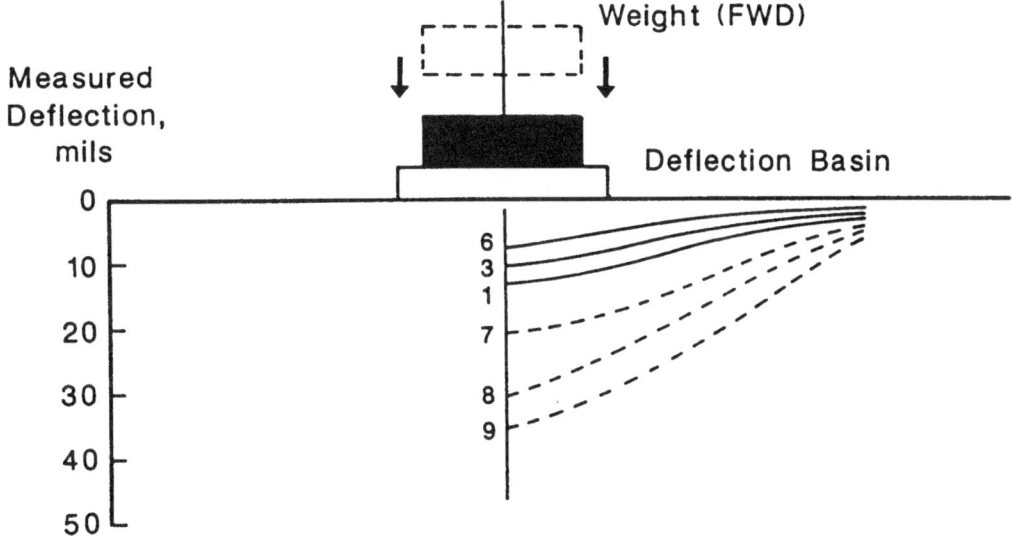

FIGURE 6.20. FWD DEFLECTION BASINS FOR SIX PHOENIX, ARIZONA, PAVEMENTS WERE SUBSTANTIALLY INFLUENCED BASED ON WHETHER OR NOT THE AGGREGATE BASE COURSE (ABC) WAS STABILIZED WITH A LOW PERCENTAGE OF LIME (AFTER LITTLE, 1990).

Table 6.5 Back-Calculated Resilient Moduli of Aggregate Base Course Layers for Arizona Pavements.

Pavement Section	Average Resilient Modulus, psi	Coefficient of Variation, percent
1 (1% Lime)	54,500	40
3 (1% Lime)	224,150	33
6 (1% Lime)	407,000	70
7 (unstabilized)	34,800	20
8 (unstabilized)	13,400	28
9 (unstabilized)	19,800	80

1 psi = 6,894 Pa

Despite the large variations, the effect of the low percentage of lime stabilization was to substantially improve the in situ stiffness of the aggregate base course layers. Each pavement section had been in place for at least one year at the time of testing.

Based on these in situ moduli, a layered elastic computer model was used to calculate the critical tensile strain within the asphalt concrete pavement layer and the vertical compressive strain at the top of the subgrade. These mechanistic parameters are widely used in pavement design and analysis protocols as discussed in Chapter 7. The results of these analyses are summarized in Table 6.6. The lower compressive subgrade strains and tensile strains in the asphalt layer result in longer performance lives for the pavements with the stabilized layers. The analysis indicates that the mode of failure for each pavement will ultimately be flexural fatigue in the surface. The expected performance lives are also summarized in Table 6.6. The lower compressive strains under the lime stabilized base is due to the superior load-spreading capability of the stabilized ABC as compared to the unstabilized ABC. The increase in flexural fatigue life of the pavement with the stabilized layers compared to the pavements without stabilization stems from the improved supporting capability of the lime stabilized base which prevents the asphalt surface from developing significant tensile strains.

A third mode of failure not accounted for in typical pavement design protocols is the potential of the asphalt concrete layer to deform due to the development of high shear stresses within the surface layer. Protection against deformation in the surface layer depends primarily on the mixture design of the asphalt concrete surface. However,

Table 6.6. Summary of Critical Strains in Phoenix, Arizona Pavements (After Little, 1990).

Pavement Identification	Base Course	Tensile Flexural Strain in HMAC, in/in x 10^{-6}	Predicted Fatigue Life (ESAL's)	Subgrade Compressive Strain, in/in x 10^{-6}	Shear Stress ratio in HMAC
Section 1	ABC (1% lime)	200	10^6	290	0.65
Section 3	ABC (1% lime)	60	2×10^7	110	0.50
Section 6	ABC (1% lime)	50	10^8	125	0.45
Section 7	ABC (unstabilized)	280	7×10^5	390	0.70
Section 8	ABC (unstabilized)	360	7×10^4	370	0.80
Section 9	ABC (unstabilized)	320	9×10^5	450	0.75

1 in. = 25.4 mm

analysis proves that better support (higher resilient modulus) provided by the aggregate base course can result in lower shearing stresses within the asphalt concrete surface. Since shear stresses induce shoving or rutting in the asphalt surface, the potential to rut can be lowered by either improving the strength or stability of the asphalt concrete surface or improving the support capability of the base or both. Table 6.6 illustrates the ratio of maximum induced shear stress within the layer to shear strength of the asphalt concrete surface at the same stress state. This is called the shear stress ratio (SSR). The potential to rut or deform decreases as the SSR decreases. The reciprocal of the SSR can be thought of as the safety factor against deformation. In this case the average safety factor against shearing-induced deformation within the surface asphalt layer is approximately 1.40 for the unstabilized sections and approximately 1.90 for the stabilized layers.

It should be noted that in place resilient moduli may be substantially lower than resilient moduli determined on intact, laboratory fabricated specimens. This is because the in situ lime stabilized pavement layer includes the effects of cracking or other forms of damage which influence the in situ response of the layer to loading.

Low percentages of lime can be used to stabilize aggregate base courses which contain reactive fines (plastic or cohesive fines). The product is usually an aggregate base course with a significantly improved resilient modulus, which has been shown to result in much improved pavement performance over the life cycle of the pavement. Low percentages of lime in aggregate base courses often result in bases with substantially higher moduli but with moduli which are substantially lower than those stabilized with higher percentages of cementitious stabilizers such as lime-fly ash or portland cement. This may have advantages under certain conditions as lime has pozzolanically stabilized the fines within the aggregate base course, yet does not develop a rigid cemented matrix. The result is a base with improved stiffness and load-carrying ability yet without the potential to shrink and crack as much as a rigidly cemented base might.

Graves et al. (1990) have also shown that lime can be used to substantially improve the compressive strength and moduli of limestone aggregate bases that do not contain clay fines. The $Ca(OH)_2$ serves as a "catalyst" to the carbonation reaction and "bonds" the aggregate system together.

In Situ Moduli and Strength Properties of Lime Stabilized Subgrades

Little (1993) reported a road test study of a lime stabilized Burleson Clay in Brazos County, Texas. The clay is a high plasticity clay (PI of approximately 40) whose optimum lime content is approximately 8 percent. The unconfined compressive strength of this clay increased from approximately 620 kPa (90 psi) to approximately 2,068 kPa (300 psi) with 48 hour curing at 49°C (120°F) when stabilized with 8 percent lime. The soil was stabilized in the road test with 5 percent lime. The unconfined compressive strength after 48 hours of curing at 49°C (120°F) of the Burleson soil stabilized with 5 percent lime was approximately 1,379 kPa (200 psi). However, the compressive strength jumped an additional 1,379 kPa (200 psi) after 28 days of moist curing at 22°C (73°F).

The FWD was used to measure in situ resilient moduli of this Burleson clay in the natural state and with 5 percent lime stabilization. All deflection testing was performed directly on the clay soil without any pavement layers over the clay layer using an FWD. Two thicknesses of stabilized clay were evaluated: 152 mm and 304 mm (6-inches and 12-inches). Measurements on the stabilized layers were made prior to and following trafficking of the pavement with 5,000 applications of a 80 kN (18,000 pound) single axle load. The results of this analysis are summarized in Table 6.7.

From this analysis several pertinent conclusions may be drawn:

1. The in situ resilient modulus of lime stabilized clay is substantially increased over the modulus of the unstabilized soil.
2. The magnitude of the stiffness increase places the load-spreading capability of the lime-stabilized clay in the stiffness category of that of good quality bases.

Table 6.7. In Place Resilient Modulus Calculation for Burleson Clay Subgrade at TTI Test Track.

Section Identification	Treatment	Depth of Treatment	Resilient Modulus, psi	
			Before Traffic	After Traffic*
8—Uncompacted Subgrade	None	—	1,000	1,000
—Compacted Clay Subgrade	None	12-in.	6,000	4,000
9—Uncompacted Subgrade	None	—	4,100	3,500
—Treated Clay Subgrade	5% Lime	12-in.	70,000	30,000
10—Uncompacted Subgrade	None	—	4,400	4,700
—Treated Clay Subgrade	5% Lime	12-in.	25,000	34,000
11—Uncompacted Subgrade	None	—	2,600	—
—Treated Clay Subgrade	5% Lime	6-in.	78,000	—
12—Uncompacted Subgrade	None	—	1,000	—
—Treated Clay Subgrade	5% Lime	6-in.	66,000	—
13—Uncompacted Subgrade	None	—	1,200	—
—Treated Clay Subgrade	5% Lime	6-in.	57,000	—
4—Untreated Subgrade	None	—	2,000	—
—Flexible Base	None	6-in.	34,000	—

* Traffic was 5,000 passes of an 18,000 axle—section was in place for two years at time of testing.

1 in = 25.4-mm

1 psi = 6,894 Pa

3. The substantial stiffness increase achieved by stabilizing the Burleson clay with 5 percent lime was maintained after exposure (without surface protection) to high rainfall levels for two years and after being trafficked by an 80 kN (18,000 pound) axle load for approximately 5,000 applications without surface protection.

One of the best ways to document the permanency of lime stabilization is through non-destructive testing of existing pavement sections. Table 6.8 documents 12 pavement sections within the State of Texas containing lime-stabilized subgrades of various types. This table also lists the thickness of each layer and the material type of each layer.

Table 6.8 also lists the calculated resilient modulus of the natural subgrade and the lime-stabilized subgrade for each pavement. It should be noted that the lime-stabilized layer in these pavements has been in service for many years.

Table 6.8. Resilient Moduli of Lime Stabilized Subgrades (LSS's) Backcalculated from Falling Weight Deflectometer (FWD) Data (After Nowlin et al., 1992).

		In Place Resilient Modulus, psi	
Highway	Pavement Section	Natural Subgrade	Lime Treated Subgrade
IH40	10-in. HMAC 15-in. ABC 14.5-in. LSS Clay Sand	13,000	92,000
SH105	2-in. HMAC 9.6-in. ABC 6.5-in. LSS Clay Sand	19,000	260,000
US77	7.5-in. HMAC 12-in. ABC 6-in. LSS Silt	12,000	430,000
SH19	11-in. HMAC 6-in. ABC 8-in. LSS Sandy Clay	15,000	160,000
SH23	3-in. HMAC 18-in. ABC 8-in. LSS clay Sand	18,000	110,000

Table 6.8. Resilient Moduli of Lime Stabilized Subgrades (LSS's) Backcalculated from Falling Weight Deflectometer (FWD) Data (After Nowlin et al., 1992). (Continued)

Highway	Pavement Section	In Place Resilient Modulus, psi	
		Natural Subgrade	Lime Treated Subgrade
SH21	8.5-in. HMAC 11-in. ABC 4.5-in. LSS Clay Sand	18,000	800,000
IH37	7-in. HMAC 10-in. ABC 6-in. LSS Sandy Clay	25,000	930,000
SH19	2-in. HMAC 11-in. ABC 9-in. LSS Sandy Clay	25,000	300,000
IH40	10-in. HMAC 15-in. ABC 14.5-in. LSS Clay Sand	13,000	92,000
US83	10-in. HMAC 10.5-in. Soil-Aggregate 5.5-in. LSS Clay	13,000	285,000
US77	2-in. HMAC 11-in. ABC 7-in. LT-Sand Sand	15,000	14,000
US59	2-in. HMAC 8-in. ABC 9-in. LS Clay Sand	10,000	35,000
IH37	10-in. HMAC 17-in. Soil-Aggregate 6-in. LSS Sandy Clay	26,000	133,000

1 in = 25.4-mm
1 psi = 6,894 Pa

The in situ modulus for the lime stabilized subbase in these pavement sections ranges from a low of 96,516 kPa (14,000 psi) for the US 77 low PI silty soil to a high of 6,570 MPa (953,000 psi) for the sandy clay stabilized with 3 percent lime for State Highway 19. Among the pavement sections, the average effect of lime was to increase the stiffness of the stabilized soil over the natural soil by a factor of approximately 20. This stiffness increase is highly variable and highly soil dependent. However, the consistent and substantial resilient modulus increase upon lime stabilization of the native subgrade among the sections is evidence of the durability and structural significance of lime stabilization.

The Dynamic Cone Penetrometer was used as a tool to ascertain in situ properties of lime stabilized layers on two Texas highways (FM 2818 and FM 3478). The DCP evaluates shear strength of the soil layer being evaluated as a function of resistance to penetration. The DCP penetration measurement has been correlated to the in situ CBR.

Figures 6.21 and 6.22 illustrate in situ shear strength improvement provided by lime stabilization, throughout the layer, in terms of in situ CBR as approximated from DCP data. Note the substantial shear strength improvement of the lime stabilized layer as compared to the natural (unstabilized) subgrade soil layer.

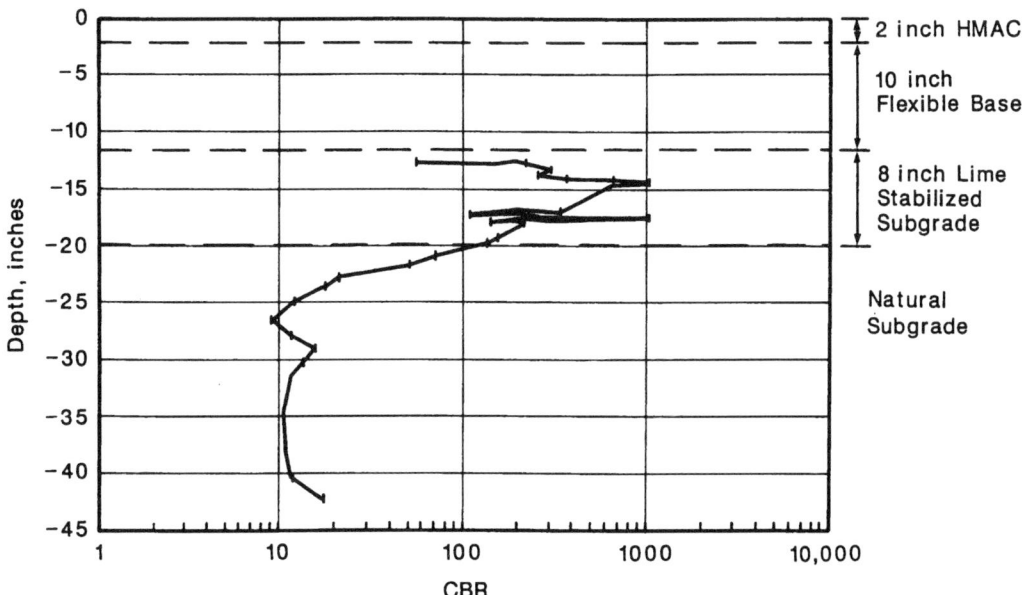

FIGURE 6.21. VARIATION IN CBR, AS APPROXIMATED BY DCP, WITH DEPTH FOR LIME-STABILIZED AND NATURAL SUBGRADE FOR FM 2818 NEAR BRYAN, TEXAS (AFTER SCULLION AND LITTLE, 1993).

1 inch = 25.4 mm

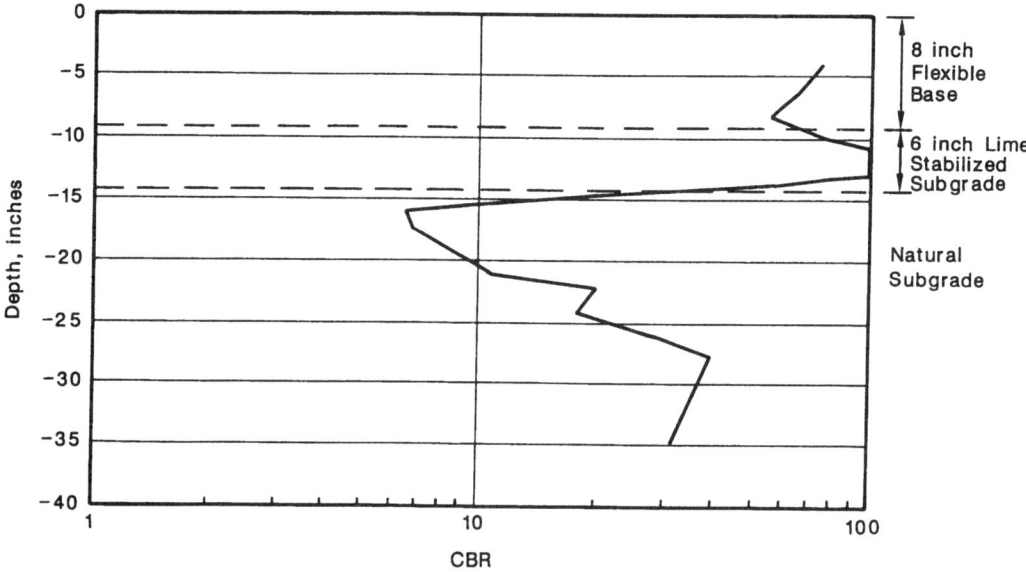

FIGURE 6.22. VARIATION IN CBR WITH DEPTH FOR FLEXIBLE HARD, LIME-STABILIZED SUBGRADE AND NATURAL SUBGRADE FOR FM 3478 (AFTER SCULLION AND LITTLE, 1993).
1 inch = 25.4 mm

Application of In Situ Moduli to Pavement Design

Probably the most widely used pavement design tool is the American Association of State Highway and Transportation Official's (AASHTO's) 1986 Pavement Design Guide. The performance equation upon which this design is based is:

$$\log_{10} N_{18} = Z_R * S_o + 9.36 * \log_{10}(SN+1) - 0.20 + \log_{10} \frac{\left[\frac{\Delta PSI}{4.2 - 1.5}\right]}{0.40 + \frac{1094}{(SN+1)^{5.19}}} + 2.32 * \log_{10} M_R - 8.07 \quad (6.2)$$

The performance equation is written in terms of applications of a 80 kN (18,000 pound) single axle load.

Notice that the decision parameters in terms of design input are the level of design reliability in terms of the normal standard deviate, Z_R; variability of design parameters and traffic (pooled standard deviation), S_o; projected traffic (in terms of the design axle load—in this case 80 kN (18,000 pound) single axle), N_{18}; acceptable level of serviceability decrease, Δ PSI, structural number, SN; and average annual roadbed modulus M_r.

Guidance for analyzing and determining these input values are presented in the 1986 AASHTO Pavement Design Guide. The only parameter that is directly related to the pavement structural design and thickness is the structural number, SN, which is defined as follows:

$$SN = a_1D_1 + m_2a_2D_2 + m_3a_3D_3$$

where a_1, a_2 and a_3 represent the structural layer coefficient of the surface hot mix asphalt concrete, aggregate base course and subbase layers, respectively; m_2 and m_3 represents the drainage coefficients of the respective granular layers and D_1, D_2 and D_3 represent the thicknesses of each respective layer.

Of these parameters, it is the structural layer coefficient, a_i, that is used to assign structural benefit to the layer. In the 1986 Design Guide the a_1 for the surface asphalt concrete is derived as a function of the dynamic modulus of the asphalt concrete mixture at 20°C (68°F). Correspondingly, the a_2 and a_3 values are also related to the resilient moduli (ASTM T 274) of the materials used for these layers. The AASHTO Guide uses the following relationships to approximate structural layer coefficients for granular layers:

$$a_2 = 0.249 \text{ (Log } E_{BS}) - 0.977 \tag{6.3}$$

where E_{BS} is the resilient modulus of the aggregate base course and

$$a_3 = 0.277 \text{ (Log } E_{SB}) - 0.839 \tag{6.4}$$

where E_{SB} is the resilient modulus of the subbase course.

Although these equations are meant to be used for determination of structural layer coefficients for unstabilized, granular base layers by AASHTO, it appears reasonable that if a low level of lime treatment can improve the resilient modulus of an aggregate base material without transforming the aggregate layer into a rigid layer, then the improvement in modulus should be reflected in an increase in the structural layer coefficient. The increase in structural layer coefficient will increase the structural number and result in a longer pavement life, with all other things being equal.

In an effort to develop structural layer coefficients for lime stabilized soils used as subbase layers for use in the AASHTO Design Guide, Thompson (1970) studied the performance of lime stabilized pavement layers in Illinois. Thompson's conclusion was that, in general, lime-stabilized soils provide structural performance equivalent to gravel or crushed stone bases. But in order to achieve this performance level the minimum unconfined compressive strength of the lime-stabilized layer must be at least 689 kPa (100 psi) after 48 hours of curing at 49°C (120°F) in areas where the destructive effects of freeze-thaw are not significant, and at least 1,034 kPa (150 psi) following accelerated cure in areas where freeze-thaw effects must be considered.

Figure 6.23 illustrates the rationale used by Thompson to determine the a_2 value for

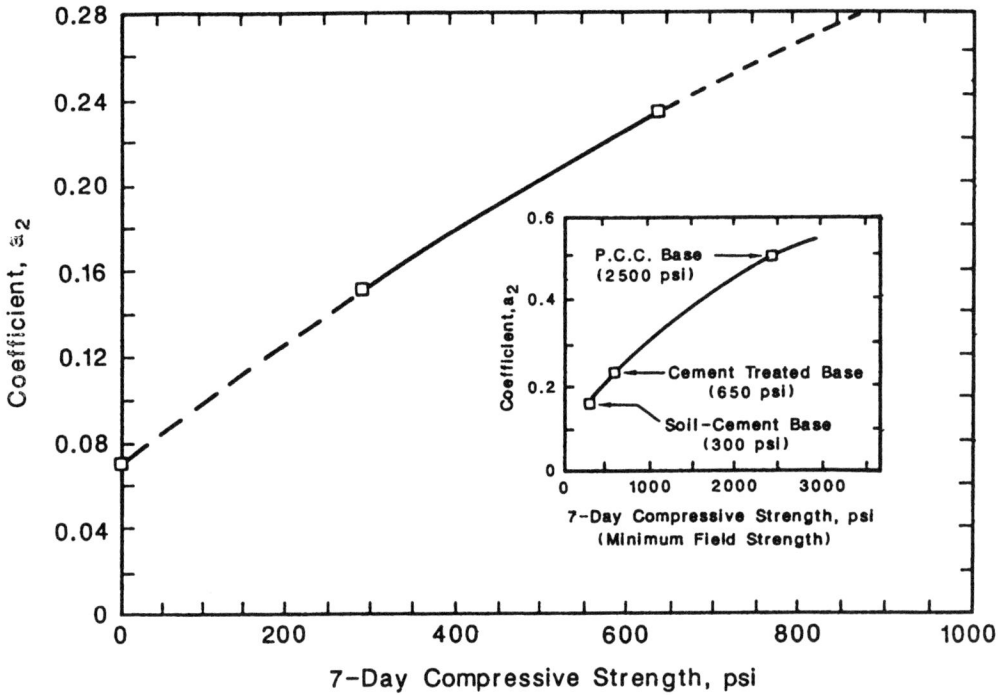

FIGURE 6.23. STRUCTURAL LAYER COEFFICIENT, a_2, WAS DETERMINED BY THOMPSON AS A FUNCTION OF COMPRESSIVE STRENGTH FOR LIME STABILIZED LAYERS (AFTER THOMPSON, 1970).
1 psi = 6, 894 Pa

lime stabilized sublayers. This figure shows the relationship between a_2 and the seven-day compressive strength of cement-treated base courses. Lime-stabilized and cement-stabilized base courses are similar with the major difference being that lime-stabilized soils gain strength at much slower rates and over longer periods of time. Thus it would appear reasonable to expect the relationship between a_2 and compressive strength to be similar to that shown in Figure 6.23, and that the unconfined compressive strength of cement-stabilized soils would compare favorably with the compressive strength of lime-stabilized soils at the onset of cold weather during the first winter following construction (estimated from samples cured for 48 hours at 49°C (120°F)). Assuming this, the value of a_2 from Figure 6.23 for a compressive strength of 1,034 kPa (150 psi) is 0.11.

Thompson further suggested that, based on the higher level of lime reactivity of certain soils which would result in unconfined compressive strengths well in excess of 1,034 kPa (150 psi), the a_2 for lime stabilized soils *could* range from between about 0.095 to 0.26 for compressive strengths of 689 to 2,758 kPa (100 to 400 psi).

The value range of the structural layer coefficient recommended by Thompson (0.09 to 0.11) for lime stabilized soils is a reasonable range considering the in situ moduli measured by Little (1990) and others. Based on Little's research and Thompson's research, it is reasonable to expect in situ moduli of between about 137,880 kPa (20,000 psi) and 482,580 kPa (70,000 psi) for fine-grained soils with unconfined compressive strengths of between 689 and 1,034 kPa (100 and 150 psi). Using equations 6.3 and 6.4 for a_2 and a_3, this equates to an a_2 of at least 0.11. With the higher in situ moduli presented in Table 6.8, the layer coefficients may be higher.

Various structural layer coefficients have been proposed in the literature for lime stabilized bases and subbases. These values typically run between 0.11 and 0.30. Van Til et al. (1972) reported AASHTO structural layer coefficients for various lime-treated bases and subbases (Table 6.9).

It is important to remember that the structural benefit of troublesome native soils *can be* substantially improved through lime stabilization. However, it is erroneous to assume that troublesome soils which are lime stabilized to improve them structurally can routinely replace high quality granular bases simply because their stiffnesses may approach those of granular bases. Both the granular base and the stabilized subbase play an important role in the performance of the structural pavement.

In addition to structural enhancement of the layer being stabilized, lime stabilization of marginal subgrades also offers the potential to improve the structural response of the aggregate base course which may rest on top of the stabilized subgrade. This effect is due to the fact that the resilient modulus response of granular base courses are

Table 6.9. AASHTO Structural Layer Coefficients Derived For Louisiana Base and Subbase Layers (After Van Til et al., 1972).

Material	Texas Triaxial Strength Class	Layer Coefficient
Base (Lime-Stabilized)		
Sand Shell	1.0	0.14
Sand-Clay Gravel	2.0	0.12
Iron Ore, Grade B	2.2	0.11
Subbase (Lime-Stabilized)		
Sand Shell	1.0	0.15
Clay Gravel	2.0	0.14
Treated Soil	3.5	0.11

dependent on the stress state developed within the layer in situ. Since the support offered by the roadbed has an important effect on the stress state developed within the aggregate base layer, one would expect lime stabilization to enhance the resilient modulus response of the aggregate base by providing better support for the base. This effect is illustrated in Tables 6.10 and 6.11 taken directly form the 1986 AASHTO Design Guide. Note that as subgrade support is improved, the bulk stress within the layer is increased and the resilient modulus response of the layer is improved substantially.

Table 6.10. Estimate of Bulk Stress (Θ) Induced in Aggregate Base Course as Influenced by Subgrade Support and HMAC Thickness (After AASHTO Pavement Design Guide, 1986).

HMAC Thickness, in	Bulk Stress (Θ), psi for Subgrade Soil with Resilient Modulus, psi of		
	3,000	*7,500*	*15,000*
Less than 2	20	25	30
2–4	10	15	20
4–6	5	10	15
Greater than 6	5	5	5

1 in = 25.4 mm
1 psi = 6,894 Pa

Table 6.11. Prediction of Aggregate Base Course Resilient Moduli Based on Moisture Content and Stress State (After AASHTO Pavement Design Guide, 1986).

Moisture State	Resilient Modulus Equation	Stress State (Θ), psi			
		$\Theta = 5$	$\Theta = 10$	$\Theta = 20$	$\Theta = 30$
Dry	$8000\,\Theta^{0.6}$	21,012	31,848	48,273	61,569
Damp	$4000\,\Theta^{0.6}$	10,506	15,924	24,136	30,784
Wet	$3200\,\Theta^{0.6}$	8,404	12,739	19,309	24,627

1 psi = 6,894 Pa

6.03 Long-term Strength of Lime Stabilized Mixtures

As has been previously discussed, lime-soil mixtures develop strength at a substantially slower rate than do portland cement stabilized base and subbase layers. In fact it is difficult to accurately predict the long-term strength of lime soil mixtures based on accelerated cure strengths. Doty and Alexander (1978) studied the strength gain of 12 California soils which were all reactive with lime. The study continued over a one year period. The physical properties of the soils and the long-terms strengths are summarized in Table 6.3.

Several important points are gleaned from these data:

1. Long-term strengths can be several times as high as the 28-day strengths, but in order for this to occur adequate lime must be used to allow full pozzolanic strength gain.
2. The advantage of slower strength gain and long-term continued strength gain should be utilized in lime stabilized mixtures. Measures should be taken to insure adequate strength at the end of the initial construction season and prior to the first Winter-Spring season.
3. The substantial long-term strength gain can result in autogenous healing of well designed mixtures so that the mixture can continually resist fatigue damage and other forms of distress which occur in periods of low level or no pozzolanic reactivity by re-establishing the pozzolanic reaction during the higher temperature curing periods.

A substantial body of field strength and physical property data exists to further substantiate the long-term durability and permanency of stabilization in lime-soil and lime-aggregate mixtures.

Lime was used to stabilize a highly expansive Taylor formation calcium-montmorillonite in 1954 in an effort to prevent undercutting during construction of IH-20 near Dallas, Texas. The addition of 4 percent lime to the raw soil reduced plasticity and increased strength as a function of time as shown in Table 6.12.

Table 6.12. Properties of Lime Stabilized Subgrade Below IH-20 Near Dallas, Texas, Illustrating Durability and Long-term Strength Development.

	Test Data	PI	Compressive Strength psi	Triaxial Classification
Raw Soil	1954	53	8	5.9*
Soil with 4% Lime	1955	26	90	1.0
Field Core Sample	1962	10	185	1.0
Field Core Sample	1964	9	195	1.0

* A triaxial classification of 5.9 represents a very poor quality subgrade while a triaxial classification of 1.0 represents a very high quality base (the highest rating).

1 psi = 6,894 Pa

The improvement of the lime-treated calcium montmorillonite was immediate and a moisture resistant, low PI soil was produced. The effect of lime stabilization was even more evident with time as strength gain continued even after seven years.

Dawson and McDowell (1961) reported similar long-term strength improvements near Round Rock, Texas, in a clayey gravel base material stabilized with 3 percent lime. The unconfined compressive strength of the raw gravel was less than 345 kPa (50 psi). After 20 days of ambient curing, the compressive strength increased to between a minimum value of 1,379 kPa (200 psi) and a maximum value of 2,413 kPa (350 psi). The compressive strength after 14 years ranged from a low value of 2,068 kPa (300 psi) to a maximum value of 4,136 kPa (600 psi).

Perhaps the most complete single field study of lime-stabilized pavement layers was reported by Aufmuth (1970). In this study, Aufmuth compared in situ CBR values of naturally occurring and lime stabilized soils. Aufmuth concluded from his evaluation that the strengths of field stabilized layers were significantly greater than the strengths of the same soils without stabilization. He further concluded that, "Based on field and laboratory CBR results, the strength derived from stabilized layers becomes permanent with age." The results of Aufmuth's study are summarized in Table 6.13.

McDonald (1969) began a study of over 1,935 Km (1,200 miles) from South Dakota pavements in 1969 and continued the study through 1983. After 13 years it was found that the lime treated soils and bases had significantly higher strengths than the untreated materials.

Table 6.14 summarizes field CBR and plate loading tests as well as laboratory CBR tests and swell tests performed on undisturbed cores from the actual pavements. The materials represented in Table 6.14 are subgrade soils from South Dakota Highway 47. This table reveals the continued good performance of the lime treated subgrade over a period of approximately 11 years.

Table 6.15 summarizes a considerable amount of data from the same McDonald study. This table compares pavements with treated and untreated subgrades and bases. The pavements are compared based on the pavement deflection measured with a Dynaflect deflection device and presented in mm; rideability, a measure of smoothness of ride (5-best to 0-worst) and maintenance cost in dollars per mile per year.

The secondary pavements compared in the tables represent the average values of 237 Km (147 miles) of pavements with untreated subgrades and 201 Km (125 miles) of pavements with treated subgrades. The last major maintenance for these pavements occurred approximately 5-8 years prior to the computation of maintenance costs. The pavements with treated subgrades have total pavement thicknesses which are 76.2 mm (3-inches) less than the thicknesses of the pavements over natural, untreated subgrades. Despite this, pavements with the treated subgrades demonstrate substantially lower maintenance costs, improved smoothness or rideability and better structural response as indicated by deflection measurements. The same general trend is seen with both

Table 6.13. California Bearing Ratio Strength Summary for Selected Pavements. (After CERL Study, Aufmuth (1970)).

Test Site	AASHTO Classification	Age, Years	Lime Content, Percent	Field CBR Unstabilized Soil	Field CBR Lime Stabilized Soil
Pine Bluff, AR					
(Site 1)	A-4 (6)	17	4.0	13	80
(Site 2)	A-4 (6)	14	4.0	38	85
Perry Co., Mo					
(Site 1)	A-7-6 (18)	14	4.5	6	77
Ft. Hood, TX	A-5 (11)	4	4.0	9	65
Frederick Co., VA	A-6 (16)	5	7.0	14	100+
Perry Co., MO	A-7-6 (18)	14	3.6	9	19
Bergstrom AFB					
(Location 1)	A-7-5 (19)	9	4.0	7	54
(Location 2)	A-7-5 (18)			5	39
Giles Co., VA					
(Location 1)	A-7-5 (18)	6	7.0	6	46
(Location 2)	A-7-6 (20)	6	7.0	4	77
Norman Co., MN					
(Site 3)	A-7-6 (20)	3	8.1	4	74
(Site 5)	A-7-6 (20)	3	3.3	6	29
(Site 7)	A-7-6 (20)	3	7.4	5	52
(Site 7B)	A-7-6 (20)	3	7.4	5	33
(Site 8)	A-7-6 (20)	3	8.1	5	72
(Site 6B)	A-7-6 (20)	3	5.1	5	52

subgrade categories: silty clay and shale clay. The same trend also continues with each pavement category or type: low traffic, medium traffic, or high (interstate-type) traffic. The data on the medium traffic pavement presented in Table 6.15 represent approximately 171 Km (106 miles) of untreated and 274 Km (170 miles) of treated pavements. At the time these data were recorded, it had been approximately 5 to 11 years since the last major maintenance operation. The data on the IH-90, high traffic pavements represent approximately 343 Km (213 miles) of untreated pavement and 229 Km (142 miles) of treated pavement. The average age of the pavements at the time of the evaluation was approximately 17 years. The age of the pavements at the time of analysis is of course very important since it is indicative of the long-term performance of lime-treated pavement layers.

Table 6.14. Field Test Data and Laboratory Test Data for Field Cores for South Dakota Highway No. 47 (After McDonald, 1969).

Test	Date	Soil Treated	Soil Untreated
Soaked CBR	1966	22%	4%
	1968	28%	4%
Field CBR	1968	48%	8%
	1975	37%	8%
Plate Load	1968	120 psi	32 psi
	1975	133 psi	40 psi
Volume Change	1964	0.2%	6%

1 psi = 6,894 Pa

Table 6.15. Comparison of South Dakota Pavements with Treated and Untreated Layers (Either Subgrade or Base Course Layers) (After McDonald, 1969).

Pavement Type	Subgrade Soil	Material Thickness, in.		Deflection, mm		Rideability Index		Maintenance $Cost/Mo	
		Treated	Untreated	Treated	Untreated	Treated	Untreated	Treated	Untreated
Secondary (Low Traffic)	East River Silty Clay	8.0	11.0	2.4	3.1	4.6	3.7	300	800
	West River Shale Clay	7.0	9.0	2.1	2.9	3.9	3.6	1,300	1,850
Primary (Medium Traffic)	East River Silty Clay	10.5	15.5	1.4	1.9	4.2	3.2	700	1,600
	West River Shale Clay	10.0	16.0	2.0	2.3	3.9	3.4	1,050	1,150
IH-90 (High Traffic)	East River Silty Clay	13.0	17.5	0.6	0.8	3.2	2.9	750	1,300
	West River Clay Shale	16.0	24.0	0.9	0.7	4.2	4.0	700	1,150
Thin Mat (1–2 in. Pavements)	17 Projects (Various Subgrade Soils)	5 in. Treated Base	10 in. Untreated Base	1.7	2.7	—	—	120	420
Seal Costs	14 Projects (Various Subgrade Soils)	5 in Treated Base	8 in. Untreated Base	3.0	3.8	—	—	260	410

1 in = 25.4-mm

In 1956, the Nebraska Department of Roads, in conjunction with the Bureau of Public Roads, now the Federal Highway Administration, built a 369 m (12,000 foot) experimental project on U.S. 136 near Tecumseh, Nebraska (Land and Ramsey, 1959). This pavement section compared both lime stabilized bases and subgrades with the standard design (unstabilized sections). The pavement sections compared in this study are summarized in Table 6.16. The subgrade soil stabilized was a plastic reddish-brown glacial clay, AASHTO classification A-7-6(16), having the following characteristics: 81 percent passing the 200 mesh sieve, liquid limit of 47 percent and a plasticity index of 26. The soil stabilized for the base was a blend of coarse sand and Aftonian silt.

Figures 6.24 and 6.25 summarize Benkleman Beam deflection data in the outer wheel path (most critical, largest deflections) for the subgrade and base comparative pavement sections, respectively. The comparisons are presented continuously over a period of approximately 1,350 days (approximately 3.75 years). These data demonstrate the substantial and enduring structural improvement offered by lime treatment of the natural subgrade and of the base material.

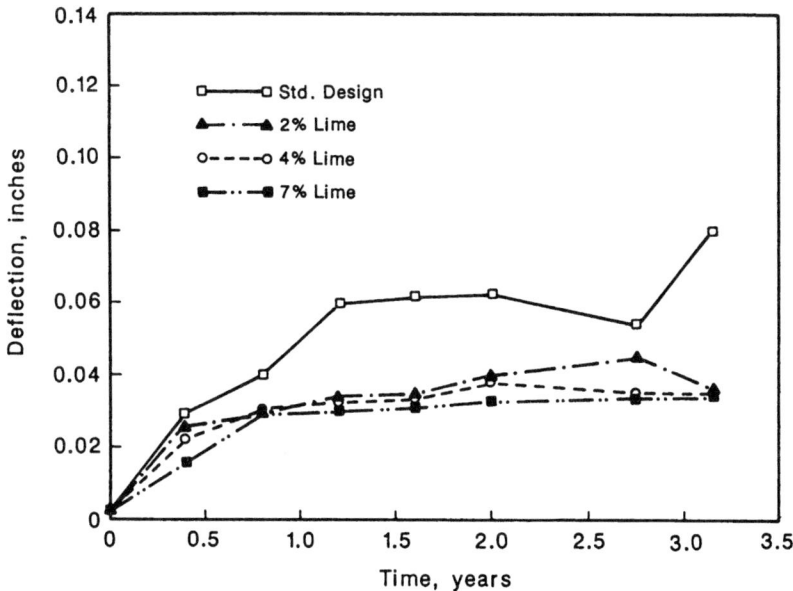

FIGURE 6.24. PAVEMENT DEFLECTIONS WERE MEASURED WITH A BENKLEMAN BEAM. SUBGRADES CONTAINED VARIOUS LEVELS OF LIME TREATMENT IN A NEBRASKA STUDY OVER A THREE YEAR PERIOD (AFTER LUND AND RAMSEY, 1959).

1 in. = 25.4 mm

Table 6.16. Summary of Nebraska Pavements Evaluated in Deflection Study of Lime Treated Base and Subgrade (After Lund and Ramsey, 1959).

Lime-Treated Subgrade Section Pavement Cross-Section Materials	Section (inches)			
	1 (Control)	2	3	4
HMAC	3-in.	3-in.	3-in.	3-in.
Soil Aggr. Base	4-in.	4-in.	4-in.	4-in.
Granular Subbase	7-in.	0	0	0
Subgrade Treatment	None	7-in.	7-in.	7-in.
Percent Lime	None	3	6	10
Total Thickness	14-in.	14-in.	14-in.	14-in.

Lime Treated Base Section Pavement Cross-Section Materials	Sections			
	5 (Control)	6	7	8
HMAC	3-in.	3-in.	3-in.	3-in.
Soil Aggr. Base	4-in.	0	0	0
Lime-Treated Bases	None	6-in.	6-in.	6-in.
Percent Lime	0	2	4	7
Granular Subbase	7-in.	5-in.	5-in.	5-in.
Total Thickness	14-in.	14-in.	14-in.	14-in.

1 in = 25.4-mm

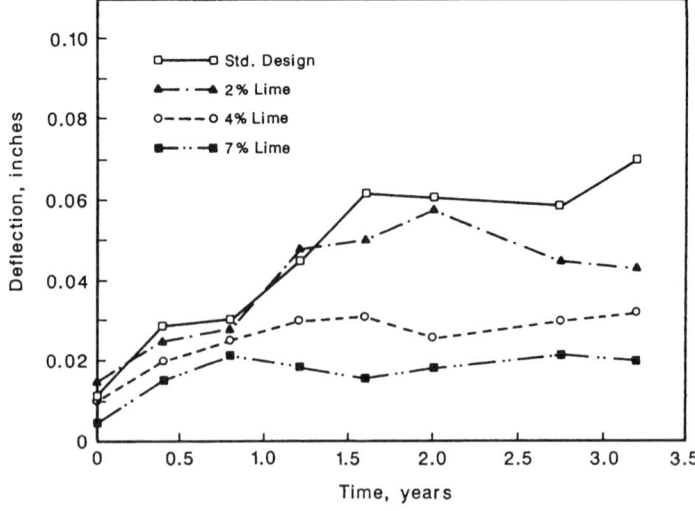

FIGURE 6.25. PAVEMENT DEFLECTIONS WERE MEASURED WITH A BENKLEMAN BEAM. BASES CONTAINED VARIOUS LEVELS OF LIME TREATMENT IN A NEBRASKA STUDY OVER A THREE YEAR PERIOD (AFTER LUND AND RAMSEY, 1959).
1 in. = 25.4 mm

6.04 Longevity of Lime Treated Soils

The longevity or permanency of lime treated soils has often been questioned. Eades and Grim (1960) questioned the permanency of lime in soil and conducted experiments with lime in pure clay minerals. They speculated that if stabilization was due to only flocculation and ion exchange, percolation of ground water could replace calcium. However, they also concluded that the formation of new compounds such as silicate and aluminate hydrates are permanent reaction products and are not susceptible to leaching. Kennedy (1988) further substantiated this conclusion by stating that once calcium silicate hydrates are formed in the stabilization process, they are permanent and do not revert. However, Kennedy did not dispute the possibility of reversal or degeneration of soil-lime effects for areas where smaller quantities of lime have been added than that which is necessary to fully stabilize the soil.

Durability of Lime Stabilization Construction Projects

Kelley's (1977) study demonstrated that soils with PI's of from 12 to 50 can be stabilized "permanently" to develop high compressive strengths (2,240 to 12,547 kPa (325 to 1,820 psi) at Fort Chaffee) and demonstrate continued strength increase with time.

The Friant-Kern Canal in California is a well documented example of durability under some of the most adverse conditions to which a stabilized layer could be subjected. In this project 4 percent quicklime was added to stabilize a PI 46 clay. The canal goes through cyclic submerged and dry conditions on a yearly basis and for two months of the year the canal is completely dry. Yet the canal is still performing well and maintaining a high resistance to erosion and a high level of slope stability.

In addition to the well documented serviceability histories of the Southwestern United States military bases and the Friant-Kern canal are numerous other construction projects, perhaps the most significant of which is the Dallas-Fort Worth International Airport (DFW). Approximately 2,045,000 square meters (2,400,000 square yards) of runways and taxiways were constructed over lime treated subgrade material at DFW. Two hundred and thirty-mm (9 inches) of lime treated soil were specified beneath taxiways and runways while 478-mm (18-inches) was specified beneath terminal aprons. According to DFW maintenance and engineering personnel, the airport has provided continuous service without major maintenance and appears that it will be able to provide good service well beyond its original design life (Long, 1989).

Leaching Effects on Soil-Lime Mixtures

Probably the greatest single concern in terms of the durability of lime-treated soils and bases is the effect of leaching on these soils. Washing or leaching of soils with permeant has the potential to react with the lime-soil mixture in one of three ways: (1) It may produce no noticeable change in the soil-water system; (2) the permeant may react with the soils to alter the soil through dissolution of chemical (cementitious) bonds, cation exchange or other processes or (3) the permeant itself may be changed without significant effects on the soil. Of these possible outcomes, the second is the one which causes concern in terms of the effect of cyclic wetting and drying on lime-soil mixtures. These cyclic wetting and drying conditions may be caused by periods of rainfall and dry periods, fluctuating water tables or periods of water runoff followed by dry periods.

McCallister-Petry Study

A comprehensive leachate study of lime-stabilized soils was conducted by McCallister and Petry (1990). In this study seven labs prepared lime treated clay samples from three different expansive soils in the North Central Texas area. The treated soils were subjected to continuous accelerated leaching for 45 and 90 days. Constants in the testing were types of soils, flow pressure, curing conditions and compaction effort. Variables were lime content, initial moisture content and duration of the leach cycle.

The major findings of the study were that:

1. The magnitude of the changes in physical and chemical property of the lime-treated soils subjected to leaching is highly dependent on the lime content of the mixture,
2. Soils stabilized with 6 to 7 percent lime demonstrated the least physical property and chemical property changes. In fact, the physical changes of the lime-treated soils at this relatively high treatment level were usually negligible and
3. Greater changes occurred at the lower stabilization rate of 3 to 4 percent lime. These changes were significant and often substantial.

The difference between lime-soil mixtures stabilized with high and low percentages of lime is probably due to the pozzolanic effect. Soils stabilized with low lime percentages often may not develop the pozzolanic reaction or at least the full complement of pozzolanic reactivity necessary to produce extensive permanent changes and resist moisture or leachate damage.

Specific chemical property changes noted in the McCallister and Petry study that substantiate the importance of using enough lime for "complete" stabilization were: (1) a significant pH decrease in the lime-stabilized soil was noticed upon leaching, but this pH decrease was found to be directly related to the diffusion of lime caused by leaching fluid thereby directly diluting the complex hydrogen compounds; (2) the amount of lime needed to offset the diffusion of lime appeared to be approximately equal to the lime content that produces the optimum level for stabilization (highest unconfined compressive strength). McCallister and Petry call this the lime stabilization optimum or LSO.

Therefore, a major conclusion of the McCallister and Petry study is that in order to provide the greatest safeguard possible against leachate damage and moisture damage in general, the proper strategy is to use the optimum lime content to produce optimum strength in the mixture design process (the LSO). This process and philosophy insures that the maximum potential for the development of pozzolanic strength gain will occur and that the "reservoir" or source of calcium ions needed to drive the pozzolanic reaction is adequate, even during periods of potential damage. This "reservoir" of calcium provides the potential for the autogenous healing effect during periods favorable for curing. This philosophy substantiates the assumptions of Eades and Grim (1960) and Kennedy (1988) that the pozzolanic reaction products are permanent.

6.05 Strength Enhancement of Lime Soil and Lime Aggregate Mixtures Through the Addition of Fly Ash and Lime Fly Ash (LFA)

General

The combination of lime and fly ash has been effective in stabilizing soils and aggregates. Fly ash is a pozzolan, which is defined by ASTM as " A siliceous or siliceous and aluminous material, which by itself possess little or no cementitious value, but will, when in a finely divided form and in the presence of moisture, chemically react with calcium hydroxide at ordinary temperature to form compounds possessing cementitious properties."

There is considerable variation in the quality and reactivity of fly ashes from different sources. In general, fly ashes from bituminous coals from the Appalachian region behave as true pozzolans, with little or no cementing property except when a source of $Ca(OH)_2$ is added. Fly ashes produced from burning coal from the mid-continent have natural setting properties because of the CaO naturally available in these ashes. Fly ashes produced from the subbituminous and lignite coals from the northern and western plains states have a high natural CaO content and may be highly cementitious, even without addition of lime.

The factors that most readily influence the quality and reactivity of fly ashes are:

1. Source of the coal,
2. Degree of pulverization of the coal and efficiency of the burning operation and
3. Collection and storage methods of the ash.

The factors that influence the extent and rate of the reaction of the lime and fly ash are:

1. Quantity of free lime added,
2. Total amount of silica and alumina in the fly ash,
3. Presence of carbon and deleterious compounds in the ash which may interfere with or inhibit the pozzolanic reaction between the lime and the silica and alumina compounds of the ash,
4. Fineness of the ash,
5. Presence of adequate moisture for reaction,
6. Compacted density of the pavement layer and
7. Temperature and age of the pavement layer.

ASTM classifies fly ashes as either type "C" or type "F" (ASTM Designation C 618). The basic difference in these two types is the percent of CaO in the ash. Procedures for evaluating the suitability of the fly ash for use in lime-fly ash mixtures are given in ASTM C 593.

Available Calcium

Despite the fact that some ashes possess high CaO contents, the CaO content reported for the ash is not all available CaO. The large majority of this CaO is chemically combined with other compounds such as silica or alumina. As such this CaO is not free to react with the soil as lime would do. Fly ashes with high CaO contents and acceptably high pozzolan contents often do indeed exhibit strong pozzolanic reactions. In fact the reaction often occurs very rapidly. The fact that this cementing reaction occurs rapidly coupled with the fact that the lime is already fused or combined with the pozzolan explains why all the CaO in the fly ash may not be available for lime—soil cation exchange and pozzolanic reactions which result in favorable soil property changes.

Even if the fly ash is reactive (type C) the addition of lime will usually enhance strength gain.

Suitable Soils and Aggregates

Virtually any granular soil can be suitably stabilized with lime—fly ash. The general mixture design approach for such stabilization is to add the amount of fly ash which will fill the voids of the mixture and provide the maximum density mixture. The next step is to add sufficient hydrated lime to maximize the pozzolanic reaction between the lime and the fly ash pozzolans. ASTM C-593 provides two criteria for judging the acceptability of lime and fly ash mixtures:

1. A minimum unconfined compressive strength following vacuum saturation of 400 psi or
2. A maximum of 14 percent weight loss following 12 cycles of freeze thaw.

Either criterion may be used to evaluate mixture acceptability.

For more plastic clayey soils the approach may be to add sufficient lime initially to the soil to reduce plasticity and improve workability and then to add sufficient fly ash to the modified mixture to support the pozzolanic strength gain. This is a rational approach when a plastic clay is to be stabilized, and lime alone will not provide the necessary pozzolanic strength gain.

Engineering Properties

The engineering properties of LFA mixtures vary considerably. The compressive strengths of LFA mixtures can be substantially higher than those of plastic clays stabilized with lime only. The strength depends on the factors listed previously which influence the rate and degree of reactions. In addition to these factors is the effect of the gradation and nature of the soil or aggregate being stabilized.

6.06 References

AASHTO, (1986). *Interim Flexible Pavement Design Guide.*

Afmuth, R. E., (1970). "Strength and Durability of Stabilized Layers Under Existing Pavements," Construction Engineering Research Lab, Report M-4.

Basma, A. A. and Tuncer, E. R., "Effect of Lime on Volume Change and Compressibility of Expansive Clays," *Transportation Research Record* No. 1295.

Dawson, R. F., and McDowell, C., (1961). "A Study of an Old Lime-Stabilized Gravel Base," *Highway Research Board,* Lime Stabilization: Properties, Mix Design, Construction Practices and Performance, Bulletin 304.

Doty, R. and Alexander, M. L., "Determination of Strength Equivalency for Design of Lime-Stabilized Roadways," Report No. FHWA-CA-TL-78-37.

Eades, J. L., and Grim, R. E., (1960). "Reactions of Hydrated Lime with Pure Clay Minerals in Soil Stabilization," *Highway Research Bulletin* No. 262.

Fossberg, P. E., (1969). "Some Deformation Characteristics of Lime-Stabilized Clay," *Highway Research Record* No. 263.

Graves, R. E., Eades, J. L. and Smith, L. L. (1990). "Calcium Hydroxide Treatment of Construction Aggregates for Improved Cementation Properties," American Society of Testing and Materials, Special Technical Publication, 1135.

Holtz, W. G., (1969). "Volume Change in Expansive Clay Soils and Control by Lime Treatment," Second International Research and Engineering Conference on Expansive Clay Soils, Texas A&M University, College Station, Texas.

Kelley, C. M., (1988). "A Long Range Durability Study of Lime Stabilized Bases at Military Posts in the Southwest," National Lime Association, Bulletin 328.

Little, D. N., Thompson, M. R., Terrel, R. L., Epps, J. A. and Barenberg, E. J., (1987). "Soil Stabilization for Roadways and Airfields," Report ESL-TR-86-19, Air Force Services and Engineering Center, Tyndall Air Force Base, Florida.

Little, D. N., (1990). "Comparison of In-Situ Resilient Moduli of Aggregate Base Courses With and Without Low Percentages of Lime Stabilization," American Society of Testing Materials, Special Technical Publication No. 1135.

Little, D. N., (1993). "Evaluation of the Structural Properties of Stabilized Pavement Layers," Interim Report to the Texas Department of Transportation, Research Project 1287.

Liu, T. K. and Thompson, M. R., (1966). "Variability of Some Selected Laboratory Tests," *Proceedings, National Conference on Statistical Quality Control Methodology in Highway and Airfield Construction,* University of Virginia, Charlottesville.

Lund, O. L., and Ramsey, W. J., (1959). "Experimental Lime Stabilization in Nebraska," *Highway Research Board,* Bulletin 231.

Maxwell, A. A., and Joseph, A. H., (1967). "Vibratory Study of Stabilized Layers of Pavement in Runway at Randolph Air Force Base," *Proceedings, Second International Conference on the Structural Design of Asphalt Pavements,* University of Michigan, Ann Arbor.

McCallister, L. D., and Petry, T. M., (1990). "Property Changes in Lime Treated Expansive Clays Under Continuous Leaching," Technical Report GL-90-17.

McDonald, E. B., (1969). "Lime Research Study—South Dakota Interstate Routes (Sixteen Projects)," South Dakota Department of Highways.

Nowlin, L., Smith, R. E., and Little, D. N., (1992). "Back Calculated In-Situ Moduli of Lime Stabilized Layers from FWD Data in Texas Pavements," Texas Transportation Institute.

Robnett, Q. L., and Thompson, M. R., (1973). "Interim Report—Resilient Properties of Subgrade Soils—Phase I—Development of Testing Procedures," Civil Engineering Studies Transportation Engineering Series No. 5, University of Illinois, Champaign—Urbana.

Seed, H. B., Woodward, R. J. and Landgren, R., (1962). "Prediction of Swelling Potential for Compacted Clays," *Journal of Soil Mechanics and Foundations Division*, ASCE, Vol 88, No. SM-7.

Scullion, T. and Little, D. N., (1993). Dynamic Cone Penetrometer Data for Project 1287, "Structural Properties of Stabilized Bases and Subbases, Texas Department of Transportation.

Suddath, L. P., and Thompson, M. R., (1975). "Load Deflection Behavior of Lime Stabilization Layers," Technical Report M-11B, Construction Research Laboratory, Champaign, Illinois.

Shrinivas, M. S., (1992). "Evaluation of the Performance of Polypropylene Fibers in Soil Stabilization," Master of Science Thesis, Texas A&M University.

Terrel, R. L, Epps, J. A., Barengerg, E. J., Mitchell, J. K. and Thompson, M. R., (1979). "Soil Stabilization in Pavement Structures—A User's Manual, Volume 1—Pavement Design and Construction Considerations," FHWA-IP-80-2, Federal Highway Administration, Washington, D. C..

Thompson, M. R., (1966). "Shear Strength and Elastic Properties of Lime-Soil Mixtures," *Transportation Research Record* No. 139.

Thompson, M. R., (19660. "Lime Reactivity of Illinois Soils," *Journal of the Soil Mechanics and Foundations Division*, ASCE.

Thompson, M. R., (1967). "Factors Influencing the Plasticity and Strength of Lime-Soil Mixtures," Bulletin 492, Engineering Experiment Station, University of Illinois.

Thompson, M. R., (1970). "Soil Stabilization for Pavement Systems—State of the Art," Technical Report, Construction Engineering Research Laboratory, Champaign, Illinois.

Thompson, M. R., and Dempsey, B. J., (1969). "Autogenous Healing of Lime Soil Mixture," *Highway Research Record* No. 263.

Townsend, D. L., and Klyn, A. T. W., (1966). "Canadian Lime Stabilization," *Rural and Urban Roads*.

Transportation Research Board, (1987). *State of the Art Report No. 5, Lime Stabilization*.

Tuncer, E. R., and Basma, A. A., (1991). "Strength and Stress-Strain Characteristics of Lime-Treated Cohesive Soil," *Transportation Research Record*, No. 1295.

CHAPTER 7

THICKNESS DESIGN CONSIDERATIONS

7.01 Pavement Design Strategies

Several different methods may be used to design flexible pavements containing lime treated pavement layers. Flexible pavement design methods usually fall into one of several categories: (1) empirical methods, (2) limiting shear failure methods, (3) limiting deflection methods, (4) regression methods based on pavement performance or road tests or (5) mechanistic—empirical methods.

Empirical Methods and Limiting Shear Failure Methods

Empirical methods such as the U. S. Army Corps of Engineer's CBR design method are usually based on some test property by which the strengths of the pavement layers are evaluated. In the case of the CBR design approach, performance data were used to empirically define the required pavement thickness over subgrade soils of various strengths. The California Bearing Ratio (CBR) is used to define the strength of the subgrade soil.

As demonstrated in Chapter 3, lime stabilization significantly improves the shear strength of lime treated subgrades, subbases and bases. Some empirical thickness design procedures, such as those based on CBR, R-value, etc., can be used to evaluate the relative performance potential of pavements with and without lime treated layers based on the relative merits of the competing paving materials as reflected by the relative strength tests. However, much work remains to be done in this area in order to develop substitution ratios for stabilized layers. Figure 7.1 is an example of the relationship between unconfined compressive strength and substitution ratios used by the Air Force (Manual 88-7, Chapter 3) in a CBR-related thickness design approach.

An example of the application of lime treated layers to structural sections in an empirical design approach is the California method based on gravel equivalency factors. Historically, the California method applied a gravel equivalency factor of 1.2 to all lime treated materials. However, because their research with lime treated soils demonstrated a wide difference in unconfined compressive strengths for various soils, it was decided that the gravel equivalency assigned to a lime treated soil should be based on the unconfined compressive strength of the mixture.

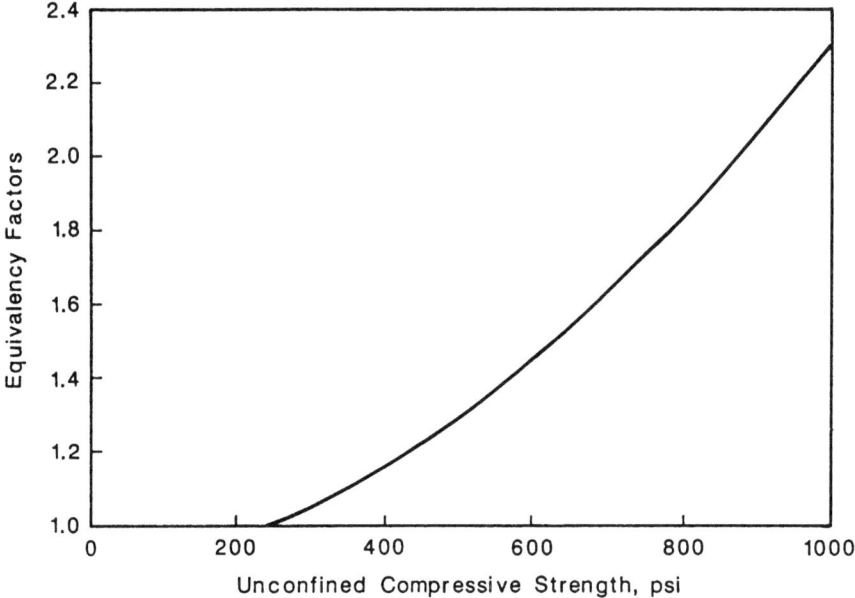

FIGURE 7.1. SOME AGENCIES, SUCH AS THE AIR FORCE MANUAL 88-7 USE EQUIVALENCY FACTORS TO DETERMINE THE THICKNESS REPLACEMENT FACTOR FOR STABILIZED LAYERS. THESE FACTORS ARE NORMALLY BASED ON STRENGTH.
1 psi = 6,894 Pa

The gravel equivalency (GE) factor for lime treated soils is calculated in the California method as:

GE (lime-treated soils) = (0.9 + UCCS/1000)

where UCCS is the unconfined compressive strength in psi after 7-days of curing at 44°C (110°F). The California method requires a minimum unconfined compressive strength of 2,758 kPa (400 psi) in order to incorporate this approach. It should be noted that the method for determining the unconfined compressive strength of the lime treated soils according to the California method is discussed in FHWA Report CA-TL-78-37 (Doty and Alexander, 1978). In that report, 12 soils from throughout California were lime treated, and the compressive strength was measured in accordance with the California procedure. Nine of the 12 soils demonstrated unconfined compressive strengths above 2,758 kPa (400 psi). The average gravel equivalent of the 12 soils stabilized with 7 percent hydrated lime was 1.4.

The Texas method of flexible pavement design employs the Texas Triaxial Test. This triaxial measure of shear strength offers an excellent test method by which to evaluate the ability of lime treatment to improve or upgrade the structural quality of a soil or aggregate material. As discussed in Chapter 6, lime treatment of marginal aggregates can improve the Texas triaxial class by one or more levels. This improvement in shear strength of the aggregate or soil through lime treatment has a significant effect on thickness design.

Limiting Deflection Methods

A popular method upon which to base pavement thickness requirements is in-place deflection measurements. Historically these in situ measurements have been made with Benkleman Beams, Dynaflects, Falling Weight Deflectometers (FWD's), etc. The general approach is that a certain pavement structural section is required for a specific subgrade and traffic level to limit surface deflections to an acceptable level.

This approach is simple and effective as it is related to in situ data. The substantial improvement in resilient modulus offered by lime stabilization of subgrade, subbase and base course layers can result in very substantial reductions in surface deflections resulting in substantially improved pavement performances.

Regression Methods Based on Pavement Performance or Road Tests

The best example of the application of regression methods from road tests is the AASHTO design procedure which is based on the results of the AASHTO Road Test. In this approach the pavement thickness is designed to support a certain number of load applications and, throughout the design life, maintain the required level of serviceability to the user.

Pavement serviceability is defined in the AASHTO approach as a function of pavement roughness, pavement rutting and pavement cracking and patching within the wheel path. The serviceability is primarily influenced by pavement roughness, secondarily by rutting and finally by cracking. The design pavement thickness is not only influenced by the level of serviceability required to be maintained but also by the acceptable level of reliability of the pavement and the average annual roadbed support as defined by the resilient modulus of the subgrade.

In the AASHTO performance equation, the contribution of the pavement structure is defined by the structural number:

$$SN = a_1 D_1 + m_2 a_2 D_2 + m_3 a_3 D_3$$

where D_1 represents the thickness of each respective layer, a_1 represents the respective structural layer coefficient of each layer, and m_1 represents the drainage coefficient of each layer. It is the structural layer coefficient that defines the relative structural contribution of a layer.

Higher values of the layer coefficient, a_1, represent a greater contribution from that layer to the performance of the pavement. Hence, in designs using the AASHTO approach, it is necessary to define the value of a_1. As discussed in Chapter 6, the value of a_2 for an unbound, granular base is substantially influenced by the support provided to that layer by the underlying subgrade. The a_2 value of the base is a function of the resilient modulus of the base.

The influence of lime stabilization of subgrade, subbase and base course layers can be assessed using the AASHTO 1986 Design Guide by evaluating the influence of the lime treatment on the structural layer coefficient of the lime-treated layers and the layers they support. Although a structural layer coefficient for a lime treated subgrade or base course layer was not developed at the AASHTO Road test, several studies have been performed to develop layer coefficients for lime treated layers. Most of these studies have placed the layer coefficient in the range of 0.095 to 0.15 with an average value of 0.11. See Section 6.02, pages 100–116.

The Illinois mixture design and pavement design with lime treated layers (Thompson, 1970) requires the development of an unconfined compressive strength of at least 100 psi for layers used as a subbase and 150 psi for lime treated layers used as a base course. According to the Illinois approach, these minimum unconfined compressive strengths relate to AASHTO structural layer coefficients of 0.12 for subbase layers and 0.11 for base course layers. Experience of the local user agency must be part of the assignment of structural layer coefficients. *Layer coefficients should not be indiscriminently applied without considering performance histories.*

A commonality exists among many of the lime-soil and lime-aggregate mixture design procedures in terms of the minimum unconfined compressive strength require-

Table 7.1. Strength Requirements for Lime Stabilized Bases, Subbases for Various State Agencies.

Agency	Strength Requirement, psi
Illinois Highway Department	100 psi (subbase) 150 psi (base)
Louisiana Department of Transportation and Development	50 psi (subbase) 100 psi (base)
Texas Department of Transportation	50 psi (subbase) 100 psi (base)
Virginia Department of Transportation	150 psi (subbase or base)

1 psi = 6,894 Pa

ments for these mixtures. Although unconfined compressive strength testing and curing methods vary among different states and agencies, many agencies require a compressive strength in the range of 689 kPa (100 psi) if the mixture is to be used as a structural paving layer. Table 7.1 is a listing of certain agency unconfined compressive strength requirements for lime-soil mixtures to be used as structural pavement layers.

In the South Dakota method of mixture design, if lime treatment improves CBR strength by 3 or 4 times that of the natural soil, then the strength and degree of stabilization of the layer is considered adequate for use as a structural layer, and the layer is assigned an AASHTO structural layer coefficient of 0.05 for soil-lime mixtures and 0.15 for base course lime mixtures (TRB Report 5, 1987).

In summary it is apparent that the lime stabilization of subgrade layers improves pavement performance by providing an additional structural layer, and by providing improved support to the aggregate base course and/or granular subbase course. This improved support allows the granular subbase and base courses to respond with a higher resilient modulus than if placed directly over an unstabilized or soft subgrade. The result is that the aggregate base responds with a higher structural layer coefficient.

Verification of the ability of stiffer subbase layers to improve the resilient modulus response of granular bases was provided by Alam and Little (1986). In their study, deflection data were used to back-calculate in-place moduli of pavement sections containing hot mix asphalt surfaces, unbound aggregate bases and subgrades stabilized with various levels of lime and fly ash (LFA). The study demonstrated that the increased stiffness of the LFA stabilized subgrade, compared to the unstabilized subgrade, increased the resilient modulus of the aggregate bases by as much as 60 percent.

In a more recent study, Little and Scullion (1993) used Falling Weight Deflectometer (FWD) data to evaluate in-place resilient moduli. In that study they compared the resilient modulus response of natural subgrade, lime stabilized subgrade and aggregate base courses on two contiguous pavement sections constructed with identical materials with the exception that one section contained a reactive lime stabilized layer and the second did not. The results are summarized in Table 7.2.

Thus the lime treated subgrade provides a "ripple" effect within the pavement that leads to improved distribution of load, better protection of the subgrade from being over-stressed and better support of the hot mix asphalt concrete surface. The protection of the subgrade, in turn reduces the potential for deep layer rutting and roughness. The improved support offered to the hot mix asphalt concrete surface may result in reduced shoving or rutting potential within the surface and reduce flexural fatigue cracking in the surface.

It is equally apparent that lime stabilization of base layers can be used to upgrade the response of the base. This upgrade is readily apparent by the increase in resilient modulus caused due to this stabilization. This modulus increase is demonstrated from

Table 7.2. In Situ Moduli Calculated from FWD Deflection Bases from SH 59 Near Houston, Texas.

Pavement Section	Pavement Layer	Back-Calculated Resilient Moduli, psi
A	12-inch Aggregate Base[1] (305-mm)	14,500
	Natural Subgrade	4,500
B	12-inch Aggregate Base[1] (305-mm)	53,900
	Lime Stabilized Subgrade	28,000
	Natural Subgrade	4,600

[1] Identical aggregate base course material.
1 psi = 6,894 Pa
1 in = 25.4-mm

both laboratory and field test data in Chapter 6. The improved resilient modulus results in an improved structural layer coefficient and hence improved performance.

Mechanistic-Empirical Methods

Mechanistic-empirical methods such as layered elastic computer models have been effectively used to calculate stresses and strains within flexible pavement systems under traffic loads as discussed in Chapter 6, and illustrated in Figure 3.1, page 22. In the layered elastic model, the pavement layers are characterized by the resilient modulus and Poisson's ratio.

As discussed in this chapter and as documented extensively in Chapter 6 of this handbook, the improvement in resilient modulus through lime treatment of subgrade and aggregate subbase and base courses can be substantial as can the improvement in performance of the pavement sections within which these lime treated layers are included. In order to determine the level of improvement in resilient modulus offered by lime treatment, it is necessary to perform a laboratory resilient modulus test in accordance with AASHTO T 274 or to derive a modulus from in situ deflection testing. However, in the absence of such test data, resilient moduli can be approximated from compressive strength tcst data as show in Table 7.3.

Table 7.3. Approximation of Resilient Modulus from Unconfined Compression Strength Data. (After Terrel et al., 1979).

Unconfined Compressive Strength, psi	Approximate Value of Resilient Modulus, psi
100–200	25,000–100,000
200–400	100,000–300,000
> 400	300,000 +

For values of unconfined compressive strength exceeding 400 psi, the resilient modulus may be approximated by the relationship:

$$E_r = 1.15 \, (UCCS) - 140 \qquad \text{(equation 7.2)}$$

where UCCS is the unconfined compressive strength in psi and E_r is calculated in ksi

1 psi = 6,894 Pa

7.02 Example of the Influence of Lime Treated Subgrade Layer in a Flexible Pavement System

One aspect of lime stabilization of subgrades often overlooked is the "ripple" effect of the improved stiffness of the lime stabilized subgrade on the overlying layers of a flexible pavement system. This section addresses this positive effect.

Figure 7.2 illustrates the effect of lime stabilization on an Illinois soil (Thompson, 1985). In Figure 7.2, the influence of freeze-thaw cycles as well as the noticeable effect of stabilization on the resilient modulus of the soil is demonstrated. In order to illustrate the influence of this level of stabilization as reflected by the resilient modulus increase, a flexible pavement structure was analyzed.

Consider a flexible pavement with 102-mm (4-inches) of HMAC and 254-mm (10-inches) of aggregate base resting on the unstabilized Tama B subgrade. The average annual resilient modulus of the HMAC is predicted to be 2,758 MPa (400,000 psi) based on the method proposed by the Asphalt Aggregate Mixture Analysis System (AAMAS) used to calculate a seasonally weighted average HMAC resilient modulus (Von Quintus et al, 1991). As illustrated by Figure 7.2, the design resilient modulus (at a deviatoric stress of approximately 35 kPa (5 psi)) for the Tama B subgrade is approximately 27,576 kPa (4,000 psi).

Table 7.4 compares the effects of lime stabilization on the resilient modulus and structural layer coefficient of two soils: Tama B and Burleson. The Burleson clay is more pozzolanically reactive with lime and producing a greater structural contribution.

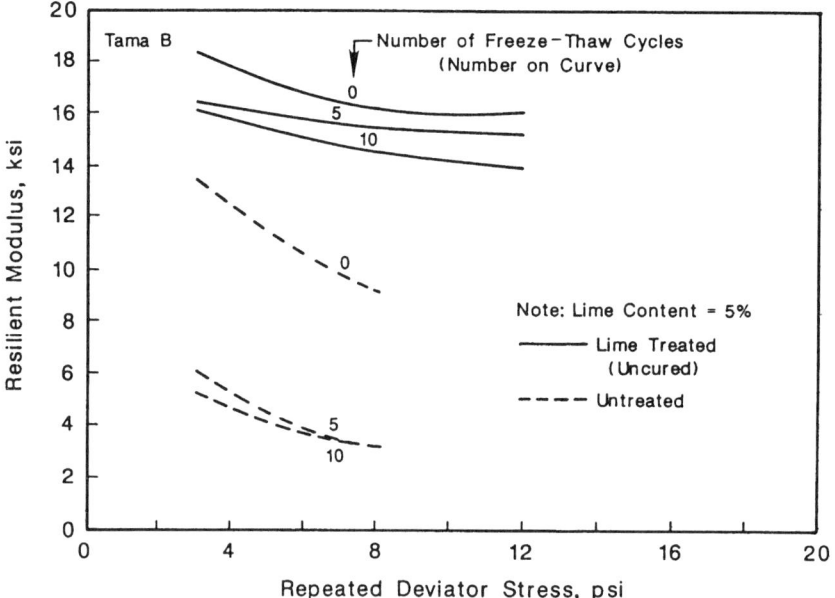

FIGURE 7.2. THE RESILIENT MODULUS OF THE TAMA B SOIL IS SIGNIFICANTLY INFLUENCED BY LIME STABILIZATION EVEN AFTER 10 FREEZE-THAW CYCLES (AFTER THOMPSON, 1985).
1 psi = 6,894 Pa

The resilient modulus of the aggregate base is highly dependent on the level of stress developed within the base layer. This point is discussed in section 3.03, page 23. In turn, the stress state is influenced by the layered pavement structure surrounding the aggregate base. This concept is employed in both the 1986 AASHTO Design Guide and The Asphalt Institute MS-1 (1982). In this analysis, two methods were used to approximate the base resilient modulus:

a. the relationship $E_{base} = 4000 \Theta^{0.6}$, where Θ is the bulk stress which is equal to the sum of the principal stresses within the layer.

b. $E_{base} = 10.447 \, (h_1^{-0.471}) \, (h_2^{-0.041}) \, (E_1^{-0.139}) \, (E_{sub}^{0.287}) \, (k_1^{0.868})$

where E_1 is the modulus of the HMAC surface, E_{sub} is the modulus of the subgrade, h_1 is the thickness of the HMAC, h_2 is the thickness of the base course and k_1 is the coefficient that appears in the resilient modulus relationship for granular bases.

In relationship (a) a layered elastic model was used to calculate the average Θ value within the base layer. A trial and error iterative approach was used to insure that the value of E_{base} used in the layered elastic analysis then produced a Θ value calculated from relationship (a) that was in reasonable agreement with the average Θ value determined in the layered elastic analysis.

Table 7.4. Comparison of Effects of Lime Stabilization on Two Clays: Tama B and Burleson, on the Resilient Modulus Response of an Overlying Aggregate Base Course (ABC).

Pavement Structures Compared	Subgrade Soil	[3]Resilient Modulus of ABC, psi	[4]a_2 of ABC	[5]Performance Life in ESAL's
A[1]	Tama B	16,000	0.07	40,000
	Burleson	16,000	0.07	40,000
B[2]	Tama B	21,000	0.09	60,000
	Burleson	28,000	0.13	110,000

[1] Pavement Structural Section: 102-mm (4-in) HMAC, 254-mm (10-in) ABC over natural subgrade
[2] Pavement Structural Section: 102-mm (4-in) HMAC, 254-mm (10-in) ABC, 204-mm (8-in) lime stabilized subgrade over natural subgrade
[3] $E_K = K_1 \Theta^{K2} = 4000 \, \Theta^{0.6}$
[4] $a_2 = 0.249 \log_{10}(E_R) - 0.977$
[5] Based on AASHTO Performance Equation
1 psi = 6,894 Pa

The value of E_{base} calculated from the Asphalt Institute's empirical relationship, (b), further illustrates the dependence of E_{base} on the pavement structure surrounding the aggregate base. The E_{base} values calculated from both approaches were in reasonable agreement.

The same analytical procedure was followed for an identical flexible pavement cross-section with the exception that the Tama B subgrade was lime stabilized for a depth of 204-mm (8-inches). The $E_{subbase}$ of this stabilized layer (from Figure 7.2) is 110,304 kPa (16,000 psi). As a result of the iterative analysis to determine E_{base} from relationship (a), E_{base} = 144,774 kPa (21,000 psi).

The comparative pavement cross-sections are summarized in Table 7.5.

Table 7.5 Comparative Pavement Structures (Pavements A and B).

	Pavement A	Pavement B
HMAC Surface: 102-mm (4-inches)	400,000 psi	400,000 psi
Aggregate Base: 254-mm (10-inches)	16,000 psi	21,000 psi
Lime-Stabilized Subgrade	None	8-inches—16,000 psi
Natural Subgrade	4,000 psi	4,000 psi

1 psi = 6,894 Pa

A mechanistic, layered-elastic analysis of pavements A and B results in calculation of the important mechanistic parameters (stresses and strains) summarized in Table 7.6.

When the values of ϵ_t, ϵ_v and τ_{max} are used in pavement analysis together with transfer functions to predict pavement distress, the performance levels summarized in Table 7.7 are predicted for each pavement.

This example illustrates substantial increases in pavement life due to the addition of lime-stabilized layers.

This illustration may be extended a step further by applying the 1986 AASHTO Design approach. In this approach, the structural layer coefficient for the aggregate base course, a_2, is significantly increased due to lime stabilization of the subgrade from 0.07 to 0.09, resulting in an increase in the structural number from pavement A to pavement B of SN_A = 2.38 to SN_B = 2.58. Disregarding any further structural contribution from the stabilized subgrade, this increase in a_2 and SN results in a pavement life improvement of approximately 50 percent (Table 7.4, column 5).

Table 7.6. Summary of Important Mechanistic Parameters Used in Predicting Pavement Performance for Pavements A and B.

Parameter	Values	
	Pavement A	Pavement B
Maximum Flexible Tensile Strain (ϵ_t) in HMAC, in./in. (Fatigue Cracking in HMAC)	0.000520	0.000418
Vertical Compressive Strain at Top of Subgrade (ϵ_v), in./in. (Deep Layer Deformation)	0.000926	0.000507
Maximum Shear Stress in HMAC Surface (τ_{max}), psi	64	54
Safety Factor Against Failure Shear Stress in HMAC	1.6	2.0

1 in. = 24.5 mm, 1 psi = 6,894 Pa

Table 7.7. Summary of Pavement Performance Prediction for Pavements A and B.

Distress Parameter	Pavement Life in Applications of an 80 kN (18,000 Pound) Axle Equivalents	
	Pavement A	Pavement B
Fatigue Cracking in HMAC (ϵ_t-Related)	126,000	214,000
Deep Layer Deformation and/or Rutting (ϵ_v Related)	45,500	730,700

Actually the benefit illustrated in the previous example as a result of the lime stabilized Tama B soil after 10 freeze thaw cycles may be conservative as much higher (than 110,304 kPa (16,000 psi)) in place moduli have been measured (Chapter 6) for lime stabilized clay soils. The effect of this can be readily seen (Table 7.4) if one considers the Burleson clay whose in-place modulus is discussed in Section 6.02, page 100. Like Tama B, the natural soil modulus of the Burleson clay is about 27,576 kPa (4,000 psi). However, the in place modulus of the lime stabilized Burleson clay is about 344,700 kPa (50,000 psi).

7.03 Example of the Role of Lime Treated Subgrade in Rigid Pavement Design

It has been established that the lime treatment of subgrades can substantially improve the modulus of subgrade reaction of these layers. This has a direct and obvious influence on the design of rigid pavement thicknesses. On many occasions, lime is used to treat expansive clay subgrades and hence reduce the volume change or swell potential of these subgrades. In addition to the reduction in volume change potential, the strength improvement and improved deformation resistance of the lime treated subgrade is essential to the performance of the rigid pavement system.

Zollinger (1989) evaluated the effect of lime treated subgrade beneath 432- mm (17-inches) of jointed PCC pavement and 229-mm (9-inches) of Portland cement stabilized base at the Dallas-Fort Worth International Airport. In this study Zollinger considered the corner loading effects of a DC-10 aircraft. The results of the analysis indicated considerable corner stresses under the DC-10 loading. These high corner compressive stresses can result in joint distress and fatigue. Failures of concrete pavements are mostly associated with the joints and how the joints perform under load. The Zollinger study calculated compressive subgrade stress as a function of bonding between the PCC slab and the cement treated base and as a function of efficiency of load transfer across the joints. The level of subgrade compressive stress as a function of these variables is summarized in Table 7.8.

Zollinger stated that subgrade stresses in excess of about 41 kPa (6 psi) for a soft clay of the type present at the Dallas-Fort Worth Airport can result in loss of support due to accumulated deformation in the subgrade. This is because stresses of this magnitude or greater can result in strain softening of these soft clays particularly when moisture contents are appreciably above optimum. This phenomenon is illustrated in Figure 7.3.

Lime stabilization can perform two critical functions in supporting a pavement com-

Table 7.8   Effect of Load Transfer Efficiency and Bonding on Subgrade Compressive Stress.

	Subgrade Stress (psi)				
	Load Transfer Efficiency				
Pavement Section	90%	75%	50%	25%	0%
17" PCC + 9" CTB Bonded	8	9	10	12.5	14.5
17" PCC + 9" CTB Unbonded	9.5	11	12	15	18

1 in = 25.4-mm
1 psi = 6,894 Pa

prised of unbonded PCC and cement-treated base especially if the potential to develop a low level of load transfer efficiency exists. The effect of lime stabilization can be to:

1. Strengthen the clay soil directly beneath the cement treated base by a factor of 10 or more and improve resistance to strain-softening and
2. Provide an additional "buffer" layer to reduce the magnitude of stress in the natural, untreated subgrade well below the critical value.

In summary, the calculated compressive stresses beneath a PCC pavement demonstrated that lime treatment of the subgrade can substantially reduce the potential to accumulate permanent deformation under heavy aircraft loadings. The result is improved performance life due to a significant reduction in the potential for slab corner deterioration.

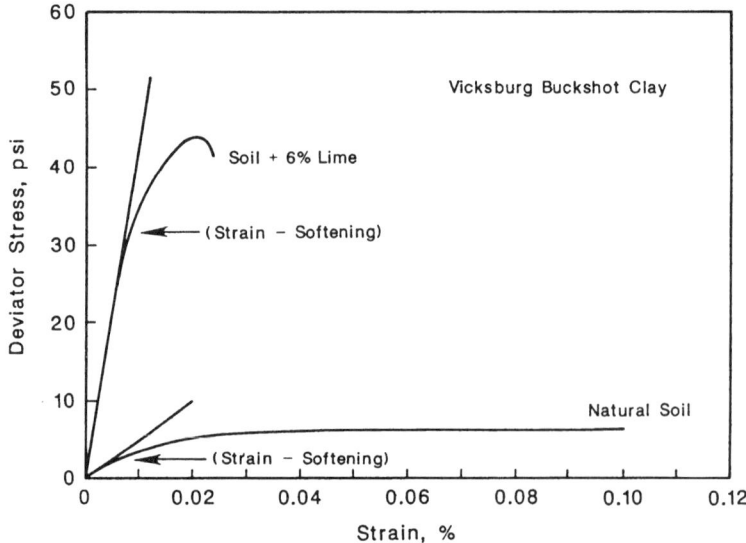

FIGURE 7.3. LIME STABILIZATION OF THE VICKSBURG CLAY SIGNIFICANTLY INCREASES THE STRENGTH OF THE SOIL AND HENCE THE STRESS LEVEL AT WHICH PERMANENT DEFORMATION WILL ACCUMULATE (AFTER TRB REPORT NO. 5, 1987).

1 psi = 6,894 Pa

7.04 References

AASHTO, (1986). *Flexible Pavement Design Guide.*

Air Force Manual 88-7, Chapter 3, *Flexible Pavements for Roads, Streets, Walks and Open Storage Areas.*

Alam, S., and Little, D. N., (1986). "Evaluation of Fly Ash and Lime-Fly Ash Test Sites Using a Simplified Elastic Theory Model and Dynaflect Measurements," *Transportation Research Record* No. 1096.

Doty, R., and Alexander, M. L., (1978). "Determination of Strength Equivalency Factors for Design of Lime-Stabilized Roadways," Report No. FHWA-CA-TL-78-37.

Little, D. N., and Scullion, T. (1993). Unpublished Data From Research Study 1287 for the Texas Department of Transportation, "Evaluation of Structural Properties of Stabilized Bases and Subbases."

Terrel, R. L., Epps, J. A., Barenberg, E. J., Mitchell, J. K., and Thompson, M. R., (1979). "Soil Stabilization in Pavement Structures, Volume I, Pavement Design and Construction Considerations," FHWA-IP-80-2, Federal Highway Administration.

Thompson, M. R., (1970). "Suggested Method of Mixture Design Procedures for Lime-Treated Soils," American Society for Testing Materials, Special Technical Publication 479.

Thompson, M. R., (1985). "Final Report—Subgrade Stability," FHWA-IL-UI-169.

Transportation Research Board, (1987). *State of the Art Report No. 5, Lime Stabilization.*

Von Quintus, H. L., Scherocman, J. A., Hughes, C. S., and Kennedy, T.W., (1991). "Asphalt Aggregate Mixture Analysis System (AAMAS)," National Cooperative Highway Research Program Report 338.

Zollinger, D., Kadiyala, S. and Smith, R. (1989), "Review of Alternate Pavement Design for Area A of Taxiway B and C Construction," Prepared for Dallas/Fort Worth International Airport, Report No. 4,100-1, Texas Transportation Institute.

CHAPTER 8

LIFE CYCLE COSTING*

8.01 Pavement Type Selection

Public agencies and private owners are constantly forced to make decisions relative to the selection of pavement type. Recognition of the factors that should be considered in the selection process and the collection and analysis of data to define these factors will help the engineers make an intelligent decision. Factors that should be considered in pavement type selection are briefly discussed below.

The United States Department of Transportation, Federal Highway Administration's policy on *pavement type selection* is stated in the October 8, 1981, issue in the Federal Register (1981). The policy states that pavement type selection should consider the following:

1. Selection should be based on an *engineering evaluation* as described by the American Association of State Highway and Transportation Officials (AASHTO, 1981).
2. Selection should be based on *life cycle cost* or economic analysis.
3. Engineering evaluation and economic analysis should be performed a *short time prior* to advertising of project.
4. *Alternate bids* should be permitted for comparable pavement designs.

The engineering evaluation as described by AASHTO (1960, 1986) should consider the following fifteen factors:

A. Principal Factors:
   1. Traffic,
   2. Soil Characteristics,
   3. Weather,
   4. Construction Considerations,
   5. Recycling and
   6. Cost Comparison.

B. Secondary Factors:
   1. Performance of similar Pavements in Area,
   2. Adjacent Existing Pavements,

---

* Sections 8.01 through 8.04 of this chapter were prepared by and are used with the consent of Dr. J. A. Epps, Dean of Engineering, University of Nevado at Reno.

3. Conservation of Materials and Energy,
4. Stage Construction,
5. Availability of Local Materials and Contractor Capabilities,
6. Traffic Safety,
7. Incorporation of Experimental Features,
8. Stimulation of Competition,
9. Municipal Preference, Participating Local Government Preference, and Recognition of Local Industry.

Thus, the selection of pavement type requires a thorough engineering evaluation and economic analysis based on detailed information. Unfortunately, these types of data are not readily available to the vast majority of public agencies.

Decisions as to pavement type on *specific projects* are routinely made (out of necessity) based on available information. Decisions as to selection of a particular pavement type for *all roadways* in a given city, county or state and/or type of highway (Interstate, Farm-to-Market, city street, etc.) are generally not made because of the absence of the needed detailed information required to justify the decision. As historic records become available from pavement management systems, these more general types of decisions as to pavement type will be possible from an engineering and economic point of view for certain groupings of pavements.

8.02 Life Cycle Cost Analysis

An economic analysis is a major part of the pavement type selection process. Life cycle costing is the conventional economic analysis tool used in pavement design.

Two methods are available to measure economic worth of pavement design alternatives, present worth (PW) or present value and equivalent uniform annual cost (EUAC). The present worth of a pavement design strategy can be viewed as the amount of money that must be available at the present time in order to have sufficient funds to pay for not only the immediate construction costs but all the anticipated future rehabilitation and maintenance operations needed through some selected period in the future.

Equivalent uniform annual costs is the amount of money required for immediate construction and all anticipated future rehabilitation and maintenance but represented as a uniform and annual expenditure. The EUAC method is often used because, alternatives with different lives can be conveniently compared and prices for alternatives are in dollar value ranges that can be easily comprehended by the lay person.

In order that the equivalent uniform annual cost or the present worth can be determined, several key items of information need to be determined and/or established.

These factors include a definition of costs, selection of a discount rate, selection of an analysis life, development of a methodology for determination of salvage value and establishment of the life of various rehabilitation alternatives. These factors are considered below.

Costs Associated With Pavement Rehabilitation

The initial and recurring costs that an agency may consider in the economic evaluation of alternative rehabilitation strategies have been defined by Haas and Hudson (1978) and include the following:
1. Agency costs:
   a. Initial capital costs of construction,
   b. Future capital costs of reconstruction or rehabilitation (overlays, seal coats, etc.),
   c. Maintenance costs, recurring throughout the analysis period,
   d. Salvage return or residual value at the end of the analysis period,
   e. Engineering and administration costs and
   f. Traffic control costs if any are involved.
2. User costs:
   a. Travel time,
   b. Vehicle operation,
   c. Accidents,
   d. Discomfort and
   e. Time delay and extra vehicle operating costs during resurfacing or major maintenance.
3. Nonuser costs.

Certainly all of these costs should be included if a detailed economic analysis is desired as described by AASHTO (1981). However, definition of many of these costs is difficult while other costs do not significantly affect the analysis of alternatives for a given pavement segment. For the sake of simplicity the method of analysis usually only considers the following costs:
1. Initial capital costs of construction,
2. Future capital costs of reconstruction and rehabilitation,
3. Maintenance costs and
4. Salvage value.

It is suggested, however, that certain user costs such as time delay costs during rehabilitation be considered on certain facilities.

Discount Rate

The discount rate selected must be based on an analytical method which is consistent in its use of either constant dollars (costs stated at price levels prevailing at a particular date in time) or current dollars (costs stated at price levels prevailing at the time the costs are incurred). A discount rate based on the market rate of return is consistent with the use of current dollars in estimating future costs. One using the real rate of return is consistent with the use of constant dollars.

The practice of using constant dollars for economic analysis together with market rate of return (current interest rate) for discounting future costs to present values is a rather common practice. However, this methodology is in error and should not be used since the market rate of return includes: 1) an allowance for expected future inflation as well as 2) a return that represents the real costs of capital. (In private investment decisions there is also included an allowance for risk; however, in Federal investments this is considered to be negligible and generally ignored.) The use of constant dollars for costing future rehabilitation and maintenance alternatives, on the other hand, makes no provision for anticipated inflation. Thus, if future costs and salvage values are calculated in constant dollars, only the real cost of capital should be represented in the discount rate used.

Constant Dollar Studies

As stated above, when constant dollar costs are used for future pavement rehabilitation and maintenance costs, the real cost of capital should be used in the analysis. The real cost of capital may be thought of as an inflation free rate of return on assets. Market interest rates approach the real cost of capital when inflation is zero. The real long term rate of return on capital has been between 3.7 and 4.4 percent since 1966 (Epps and Wootan, 1981, and Corps of Engineers, 1987). A discount rate of return of four percent is therefore suggested for analysis purposes when constant dollars are used to estimate future rehabilitation and maintenance costs and salvage value.

Current Dollar Studies

If costs are projected in inflated or current dollars, the full market rate of interest should be used. A range of eight to twelve percent has been commonly used to represent the average long-term market interest rate in recent economic studies of public projects. The United States Office of Management and Budget prescribes a ten percent discount rate for most federal government economic studies using current dollar costs (Corps of Engineers, 1987).

If current dollar costs are employed in the study, use of an average rate of inflation for all price changes is recommended unless there are good reasons to expect highly significant differences in the rate of price change for certain rehabilitation or maintenance alternatives. Inflation rates for construction and rehabilitation and maintenance materials have been higher than those experienced for consumer commodities as expressed by the Consumer Price Index over the last 15 years.

Discussion

Except for special cases where some items are expected to have significantly different rates of inflation, the consensus of economists is to use constant dollar costs and discount rates which represent the real costs of capital. In general, economists outside of government agree on this approach and cite the following primary reasons against inclusion of inflation rates in economic studies:

1. Difficulties in predicting future inflation rates,
2. The acceptance of inflation as a norm may be counter to the Government's desire for price stabilization,
3. Federal programs, if justified in part by inflating benefits, may themselves contribute to inflation,
4. Debtor's gains through repaying outstanding debts with inflated dollars are offset by creditor's losses,
5. Future dollars to pay for future expenses will likewise be inflated and therefore there is no net change and
6. A bias toward capital-intensive and long-lived projects results, making adaption to future changes more costly (Lee and Grant, 1965).

Recommendation

Comparison of pavement rehabilitation alternatives should be based on the use of constant dollars for estimating present and future costs together with salvage values. A discount rate of four percent is suggested for present value calculations.

Because the results of present value are sensitive to the discount rate, the analyst may want to perform the economic calculations at two or three alternative discount rates. It should be noted that rehabilitation alternatives with large initial costs and low maintenance or user costs are favored by low interest rates. Conversely, high interest rates favor strategies that combine low initial costs with high maintenance and user costs.

A discount rate of four percent has been utilized for examples in this handbook. Present worth factors and capital recovery factors for discount rates of 3.5, 4.0, 4.5 and 5.0 percent are shown in Table 8.1. Values for other discount rates can be found in

(Yoder and Witczak, 1976) or textbooks on engineering economy. Both present worth and the uniform annual cost methods are illustrated in the handbook. Costs are estimated in terms of dollars per square yard (1 sy = $0.92m^2$).

Table 8.1. Present Worth and Capital Recovery Factors (After Yoder and Witczak, 1976).

Years	Present Worth Factor Interest Rate %				Capital Recovery Factor Interest Rate %			
	3.5	4.0	4.5	5.0	3.5	4.0	4.5	5.0
1	0.9662	0.9615	0.9569	0.9524	1.03500	1.04000	1.04500	1.05000
2	0.9335	0.9246	0.9157	0.9070	0.52640	0.53020	0.53400	0.53780
3	0.9019	0.8890	0.8763	0.8638	0.35693	0.36035	0.36377	0.36721
4	0.8714	0.8548	0.8386	0.8227	0.27225	0.27549	0.27874	0.28201
5	0.8420	0.8219	0.8025	0.7835	0.22148	0.22463	0.22779	0.23097
6	0.8135	0.7903	0.7679	0.7462	0.18767	0.19076	0.19388	0.19702
7	0.7860	0.7599	0.7348	0.7107	0.16354	0.16661	0.16970	0.17282
8	0.7594	0.7307	0.7032	0.6768	0.14548	0.14853	0.15161	0.15472
9	0.7337	0.7026	0.6729	0.6446	0.13145	0.13449	0.13757	0.14069
10	0.7089	0.6756	0.6439	0.6139	0.12024	0.12329	0.12638	0.12950
11	0.6849	0.6496	0.6162	0.5847	0.11109	0.11415	0.11725	0.12039
12	0.6618	0.6246	0.5897	0.5568	0.10348	0.10655	0.10967	0.11283
13	0.6394	0.6006	0.5643	0.5303	0.09706	0.10014	0.10328	0.10646
14	0.6178	0.5775	0.5400	0.5051	0.09157	0.09467	0.09782	0.10102
15	0.5969	0.5553	0.5167	0.4810	0.8683	0.08994	0.09311	0.09634
16	0.5767	0.5339	0.4945	0.4581	0.08268	0.08582	0.08902	0.09227
17	0.5572	0.5134	0.4732	0.4363	0.07904	0.08220	0.08542	0.08870
18	0.5384	0.4936	0.4528	0.4155	0.07582	0.07899	0.08224	0.08555
19	0.5202	0.4746	0.4333	0.3957	0.07294	0.07614	0.07941	0.08275
20	0.5029	0.4564	0.4146	0.3769	0.07036	0.07358	0.07688	0.08024
21	0.4856	0.4388	0.3968	0.3589	0.06804	0.07128	0.07460	0.07800
22	0.4692	0.4220	0.3797	0.3418	0.06593	0.06920	0.07255	0.07597
23	0.4533	0.4057	0.3634	0.3256	0.06402	0.06731	0.07068	0.07414
24	0.4380	0.3901	0.3477	0.3101	0.06227	0.06559	0.06899	0.07247
25	0.4231	0.3751	0.3327	0.2953	0.06067	0.06401	0.06744	0.07095
26	0.4088	0.3607	0.3184	0.2812	0.05921	0.06267	0.06602	0.06956
27	0.3950	0.3468	0.3047	0.2678	0.05785	0.06124	0.06472	0.06829
28	0.3817	0.3335	0.2916	0.2551	0.05660	0.06001	0.06352	0.06712
29	0.3687	0.3207	0.2790	0.2429	0.05545	0.05888	0.06241	0.06605
30	0.3563	0.3083	0.2670	0.2314	0.05437	0.05783	0.06139	0.06505

Analysis Life

In economic studies, projects under consideration are defined as having a service life, an economic life, and an analysis life. Service life estimates the actual total usage of a facility. It is the time span from installation of a facility to retirement from service. The ending of service life of a pavement (except by disaster) is by man-made decision.

The economic life is the life in which a project is economically profitable or until the service provided by the project can be provided by another facility at lower costs. The economic life may be less than the service life. Shortage of capital often extends a project service life beyond the end of its economic life.

Analysis life may not be the same as the service life or economic life of a project, but it represents a realistic estimate to be used in economic analysis. The analysis period utilized should be long enough to include the time between major rehabilitation actions for the various rehabilitation activities under study. However, the analysis period should not be excessive as the analysis becomes more uncertain due to changes in technology and/or events not occurring as predicted. The Highway Engineering Handbook (1960). "stresses that use of an analysis life not to exceed 40 years on the basis that a sound investment should return its costs within that length of time". Suggested values to use for analysis life are shown on Table 8.2.

Table 8.2. Recommended Analysis Life for Comparing Pavement Alternatives (After Epps et al., 1987).

Activity	Pavement Surface Type	Recommended Analysis Life, Year
New Construction	PCC only	45
Reconstruction	HMA only	30
Thick Overlays	PCC and HMA	45
Rehabilitation	PCC only	20
	HMA only	20
Maintenance	PCC only	20
	HMA only	20

PCC—Portland cement concrete
HMA—hot mix asphalt

Salvage Value

Salvage value is the economic residual value of the facility at the end of the analysis period for the project. The present value of this residual value is used to partially offset the present worth of the project costs. In a broad sense, the salvage value is the remaining value of the land, equipment or facility that has continued or alternative uses at the end, or terminal year of the analysis period.

In several studies made on salvage value of pavements it was considered valid to assume zero salvage value at the end of the analysis period. However, the evaluation of pavement rehabilitation alternatives requires that some consideration be given to salvage value. The residual value of rehabilitation action based on its anticipated remaining life appears to be the best method for determining salvage value in this handbook. A simplified but adequate method is described by the equation given below:

$$SV = \left(1 - \frac{L_A}{L_E}\right) C$$

where
- SV = salvage value or residual value of construction or rehabilitation alternative,
- L_A = analysis life of the rehabilitation alternative in years i.e., difference between the year of construction or rehabilitation and the year associated with the termination of the life cycle analysis,
- L_E = expected life of the rehabilitation alternative and
- C = cost or price of rehabilitation alternative.

For example, if an analysis period of 20 years is utilized on a project where rehabilitation alternative has a life cycle of seven years, the residual or salvage value of the second rehabilitation action is equal to the straight line depreciated value of the alternative at the end of the analysis period as given by the equation above. Thus, the residual value at the 20th year would be (if the cost of the rehabilitation alternative was $2.50/sq. yd.)

$$SV = \left(1 - \frac{6}{7}\right) 2.50 = \$0.36 \; sq \; yd$$

Life of Rehabilitation Alternatives

The expected life of new construction, rehabilitation alternatives, and maintenance alternatives must be based on the engineer's experience with consideration given to local materials, environmental factors and contractor capability. Typical life cycles are shown on Tables 8.3 and 8.4.

Table 8.3. Typical Life Cycles.

Pavement Type	Representative Range	Representative Average
New PCC	16–25 yrs	20 yrs
PCC Overlay	10–20 yrs	15 yrs
New HMA	12–16 yrs	14 yrs
HMA Overlay	5–15 yrs	10 yrs

Table 8.4. Average Life Cycles (After Corps of Engineers, 1987).

Maintenance Activity	Average (mean) Life Cycle (yrs)	Life Cycle Range (yrs)	No. Data Pts.
Crack Sealing (flexible)	4	1–12.5	39
Chip Seal (flexible)	5	2–10	24
Shallow Patch (flexible)	3	0.5–10	40
Deep Patch (flexible)	6	1–20	41
Slurry Seal (flexible)	5	1–10	22
Cold Milling (flexible)	10	10	6
Heater Planing (flexible)	6	2–12.5	7
Crack Sealing (rigid)	5	1–12.5	44
Joint Sealing (rigid)	7	2–15	48
Shallow Patch (rigid)	5	0.5–25	45
Deep Patch (rigid)	8	0.55–25	36
Slab Replacement (rigid)	19	4–40	42
Grinding (rigid)	11	7.5–16	4
Mud Jacking (rigid)	16	10–30	3

* Derived from responses to a questionnaire in 1985 from forty Air Force Bases.

8.03 Cost Data

Data are available which define prices associated with pavement construction, reconstruction, recycling and maintenance operations. A large price variation can be expected depending on the location of the project and the time of construction.

The engineer should be aware that the term "pavement price" refers to the total amount of monies that an agency, or the public, must spend to have a pavement structure constructed, rehabilitated or maintained. Pavement price includes pavement costs, general contractor overhead and contractor profit. Pavement costs is defined as the amount of monies that a contractor must spend for labor, materials, equipment, sub-contracts and overhead to construct, rehabilitate, or maintain a pavement structure.

8.04 Analysis Procedure

Based on the information presented above, equivalent uniform annual cost and present worth economic evaluation methods appear to be the best suited for evaluating pavement construction, rehabilitation and maintenance strategies. A discount rate of four percent is suggested for use with an analysis period of 20 to 45 years depending upon the alternatives considered. Salvage values should be calculated based on the residual value equal to the straight-line depreciated value of the rehabilitation alternative at the end of the analysis period. The life and initial price of various rehabilitation, recycling and maintenance alternatives should be based on the engineer's experience with consideration given to local materials, environmental factors, and contractor capability. Typical price and cost data have been included for reference purposes.

The basic equation for determining present worth of rehabilitation and maintenance for a given facility is shown below:

$$PW = C - M_1 \left(\frac{1}{1+r}\right) n_1 + \ldots M\left(\frac{1}{1+r}\right) n_1 - S\left(\frac{1}{1+r}\right) z$$

where:
PW = Present worth or present value,
C = Present cost of initial rehabilitation activity,
M_1 = Cost of the 1th maintenance or rehabilitation alternative in terms of resent costs, i.e., constant dollars,
r = Discount rate (four percent suggested for use in this manual),
n_1 = Number of years from the present to the 1th maintenance or rehabilitation activity,
S = Salvage value at the end of the analysis period and
z = Length of analysis period in years (20 years suggested for use in this handbook).

The term:

$$\left(\frac{1}{1+r}\right)^n$$

is commonly called the single payment present worth factor in most engineering economic textbooks.

The present worth or present value of all costs over the analysis period can be stated in terms of EUAC by multiplying PW by the uniform series capital recovery factor:

$$\text{EUAC} = \text{PW} \times \text{crf}(r, N)$$
$$= \text{PW} \times r(1+r)^N / (1+r)^{N-1}$$

where:

PW = Present worth as before,

crf (r, N) = The uniform series capital recover factor for discount rate r and analysis period N.

Values of single payment present worth factors and uniform series capital recovery factors can be obtained from most engineering economic textbooks. Typical values for selected interest rates are shown in Table 8.1.

Table 8.5 is a calculation form for determining the present worth and equivalent uniform annual cost. The use of this form is illustrated in an example that follows. However, if the engineer anticipates the need of performing a large number of these calculations, spread sheet type personal computer programs are easily developed and greatly reduce calculation time.

8.05 Sensitivity Analysis

Present worth life cycle costs determinations are sensitive to the following factors:

1. Selected discount rate,
2. Length of analysis period,
3. Life of rehabilitation alternative,
4. Salvage value,
5. Price and cost values and
6. Consideration of user costs.

Sensitivity analyses should be conducted as part of the economic analysis. Discount rates, life cycles, analysis periods, and costs can be varied and present worth or equivalent uniform annual costs recalculated to determine sensitivity.

8.06 Example Problem

Consider the construction of two alternative pavements. The first alternative contains a 102-mm (4-inch) HMAC pavement surface underlain by a 305-mm (12-inch)

Table 8.5. Calculation Form for Present Worth Life Cycle Costing.

Year	Cost, Dollar Per Square Yard	Present Worth Factor 4 Percent	Present Worth, Dollars
Initial Cost		1.0000	
1		0.9615	
2		0.9246	
3		0.8890	
4		0.8548	
5		0.8219	
6		0.7903	
7		0.7599	
8		0.7307	
9		0.7026	
10		0.6756	
11		0.6496	
12		0.6246	
13		0.6006	
14		0.5775	
15		0.5553	
16		0.5339	
17		0.5134	
18		0.4936	
19		0.4746	
20		0.4564	
Salvage Value		0.4564	

Total = _____ Total = _____

Uniform Annual Cost = Present Worth x Capital Recovery Factor

= _____ x 0.07358

aggregate base course (ABC) over natural clay subgrade Burleson, Texas, (Table 6.7, page 108) compacted to a depth of 610-mm (24-inches) to meet 95 percent density according to AASHTO T-180. The second alternative pavement is the same as the first except that it contains an additional layer of 204-mm (8-inches) of lime stabilized clay subgrade. The pavement alternative are summarized in Table 8.6.

Table 8.6. **Example Problem Pavement Alternatives.**

Alternative 1	*Alternative 2*
Surface: 102-mm (4-inches) HMAC	Surface: 102-mm (4-inches HMAC)
Base: 305-mm (12-inches) ABC	Base: 305-mm (12-inches) ABC
Natural Clay Subgrade (Burleson Clay	Subbase: 204-mm (8-inches) Lime Stabilized Subgrade (LSS) (Burleson Clay)

1-in = 25.4-mm

A layered elastic computer analysis was performed to determine the stresses and strains developed within the pavement cross-section that are related to pavement performance. In this analysis a dual 20,000 kN (4,500 pound) wheel load was used, which represents an 80,000 kN (18,000 pound) single axle load.

In a layered elastic computer model, the layers are characterized by the elastic modulus or stiffness and the Poisson's ratio of each layer. The stiffness of an asphalt concrete layer is dependent on the temperature at the time of testing and the rate at which the load is applied to the pavement or the speed at which the traffic moves across the pavement. The greatest effect on asphalt stiffness in the field is temperature. For this analysis, the average annual stiffness of the asphalt concrete surface was determined to be 2,758 MPa (400,000 psi).

For an approximation of the elastic modulus of the ABC, the resilient modulus test was used as is explained in Chapter 6. In this test the load applied to the soil sample is a cyclic load which simulates the wheel load applied to the pavement in situ in terms of the magnitude of the stress imposed by the load, the form of the load application and the duration of the load application. As discussed in Chapter 6, the primary factors which influence the stiffness or resilient modulus of an unbound aggregate base are the stress state within the base and the moisture content and state of drainage of the base. For a damp base course (see Chapter 6), the base resilient modulus can be approximated as:

$$E_{BS} = K_1 (\Theta)^{K_2}$$

where K_1 and K_2 are regression constants, and Θ is the bulk stress invariant or a measure of the state of stress in the ABC.

For a damp base course the relationship used for determination of the resilient modulus of the ABC was $E_{BS} = 4{,}000 \, (\Theta)^{0.6}$. The average bulk stress within the ABC layer in pavement alternative 2 is considerably higher than the average bulk stress in alternative 1. Thus based on the relationship between E_{BS} and Θ, the resilient modulus of the ABC for alternative 1 is approximately 103,480 kPa (15,000 psi) and for the ABC for alternative 2 the resilient modulus is approximately 179,244 kPa (26,000 psi).

The doubling of the modulus of the ABC is due to improved support of this granular layer by the LSS. This improved support is provided in this case because the lime stabilized clay develops a response modulus of approximately 448,110 kPa (65,000 psi) as an annual average whereas the average roadbed modulus of the native clay is approximately 24,129 kPa (3,500 psi). The moduli of the various layers used in the layered elastic design are summarized in Table 8.7.

Table 8.7. Modulus Values for Pavement Layers Used in Example Problem Pavement Alternatives.

Alternative 1	Alternative 2
HMAC: 2,758 MPa (400,000 psi)	HMAC: 2,758 MPa (400,000 psi)
ABC: 103,480 KPa (15,000 psi)	ABC: 179,244 KPa (26,000 psi)
Native Subgrade: 24,129 KPa (3,500 psi)	Stabilized Subgrade: 448,110 KPa (65,000 psi)
	Native Subgrade: 24,129 KPa (3,500 psi)

1 psi = 6,894 Pa

A summary of the mechanistic parameters most often used to compute pavement distress for the two alternatives is presented in Table 8.8. The two parameters most commonly used in pavement design are the maximum tensile strain in the asphalt layer and the vertical compressive strain at the top of the subgrade layer. The HMAC tensile strain is associated with cracking in that layer, and the vertical compressive strain at the top of the subgrade is associated with deep-layer rutting and pavement roughness. As can be seen from Table 8.8, the incorporation of the lime stabilized layer considerably reduces the magnitude of both design strains.

Table 8.8. Summary of Critical Design Parameters Used in Life Cycle Example.

Tensile Strain (ϵ_t) in HMAC, in/in	Alternative 1			Alternative 2
	As Constructed	1st Overlay	2nd Overlay	As Constructed
Tensile Strain (ϵ_t) in HMAC, in/in	0.00585	0.000360	0.000260	0.000343
Fatigue Life, Years	5	21	67	30*
Subgrade Compressive (ϵ_v) strain, in/in	0.002130	0.000850	0.000580	0.000485
Predicted Life, Years Based on ϵ_v	1*	6*	30*	59
Performance Period Based on AASHTO Analysis, Years	1	8	33	34

* Indicates Critical Parameter and Critical Distress Mode
1 in. = 25.4 mm

Pavement fracture fatigue life and pavement rutting deterioration life were computed based on two well known transfer functions. The Finn (1977) relationship was used to predict fatigue cracking,

$$\log N_f = 15.947 - 3.291 \log (\epsilon_t) - 0.854 \log E_{AC}$$

where N_f is the number of load applications until 10 percent fatigue cracking occurs in the wheel path, ϵ_t is the tensile strain in the HMAC in micro in./in., and E_{AC} is the asphalt concrete stiffness in ksi. The Shell model (Claessen et al., 1977) was used to calculate life based on the criterion of rutting and/or pavement roughness:

$$N_f = 1.365 \times 10^{-9} (\epsilon_v)^{-4.447}$$

where N_f is the number of design load applications until rutting failure, and ϵ_v is the vertical compressive strain at the top of the subgrade.

Based on this analysis for 10 percent cracking and rutting, rutting was found to control in each case for alternative 1. If alternative 1 is selected, two overlays in some form of staged construction are required within the first five or six years in order to allow the pavement to withstand the structural demand imposed by application of 40 design axle equivalents per day. If the lime stabilized layer is included (alternative 2), then staged construction is not required in order to develop adequate structural integrity based on the calculated mechanistic parameters of ϵ_v and ϵ_t and the transfer functions previously discussed.

Table 8.9 presents the calculation form for present worth life cycle costing. The times when overlays and other rehabilitation and maintenance strategies are required based on the mechanistic analysis are shown in Table 8.9.

The analysis period for this example is 30 years. The present worth comparisons at the end of the 30 year period for alternatives 1 and 2 are $24.49 and $22.45, respectively. This demonstrates an 8 percent saving by using alternative 2 based on a 30 year analysis period and the assumptions incorporated in this example.

The 1986 AASHTO flexible pavement design guide can also be used to calculate changes in serviceability and when rehabilitation in the form of structural overlay is required. If this approach is used to evaluate alternatives 1 and 2 in this example, it leads to approximately the same solution as the previously discussed mechanistic ap-

Table 8.9. Calculation of Percent Worth Life Cycle Costing.

Year	Cost, Dollars per Square Yard		Present Worth Factor, 4%	Present Worth Dollars	
	Alternative 1	Alternative 2		Alternative 1	Alternative 2
Initial	$17.64	$21.64	1.000	$17.64	$21.64
2	3.72 (Overlay)	—	0.9246	3.44	—
6	3.72 (Overlay)	—	0.7903	2.94	—
8	—	0.50 (Chip Seal)	0.7307	—	0.36
16	0.50 (Chip Seal)	0.50 (Chip Seal)	0.5339	0.27	0.27
23	0.50 (Chip Seal)	0.50 (Chip Seal)	0.4057	0.20	0.20
30	3.72 (Overlay)	3.72 (Overlay)	0.3087	1.15	1.15
Salvage Value	3.72	3.72	0.3087	1.15	1.15
		TOTAL PW Dollars		$24.49	$22.47

Alternative 2 represents an 8 percent savings.
154 = 0.92 m²

proach. Using the AASHTO approach and a 30 year analysis period, alternative 2 yields approximately an 8 percent savings over the analysis period of 30 years.

In the AASHTO analysis the original pavement (alternative 1) requires an overlay at the 2 year point in order to prevent deterioration of the serviceability below the 2.5 level. A second overlay would be required at approximately 8 years and a third after approximately 30 years (at the end of the analysis period). These projected times for overlay are in close agreement with the mechanistic analysis approach. Alternative 2 requires one overlay at approximately 30 years using either the mechanistic or AASHTO analysis approach.

In the AASHTO analysis, structural layer coefficients for the ABC were calculated as a function of the resilient modulus of the ABC in accordance with the AASHTO structural layer coefficient—resilient modulus relationship (see chapter 6). The structural layer coefficient used for the lime stabilized layer (a_3) was 0.10.

It should be noted that in this analysis only the cost of the maintenance and rehabilitation activities are considered. The user utility costs are also an important factor but were not considered in this analysis. The utility costs of alternative 1 will be considerably higher than for alternative 2 during the 30 year analysis period because of the two additional major rehabilitation procedures required in years 2 and 6 for pavement alternative 1 that are not required for pavement alternative 2.

This life cycle example is for illustrative purposes only. It is not an actual design nor evaluation. However, it does illustrate the use of sound engineering considerations and life cycle comparisons. It must be understood that life cycle comparisons are sensitive to the long-term maintenance and rehabilitation strategies selected.

8.07 References

AASHTO, (1981). "Pavement Determination and Documentation, an International Guide on Project Procedures."

AASHTO (1986). "Guide for Design of Pavement Structures."

Claessen, A. I. M., Edwards, J. M., Sommer, P., and Uge, P., (1977). "Asphalt Pavement Design: The Shell Method," *Proceedings, Fourth International Conference on Structural Design of Asphalt Concrete Pavements,* University of Michigan, Ann Arbor.

Corps of Engineers, (1987). "Life Cycle Costs for Pavements," Report F84-63.

Epps, J. A., and Wootan, C. V., (1981). "Economic Analysis of Airport Rehabilitation Alternatives—An Engineering Manual," Report DOT/FAA/FD-81/78, Federal Aviation Administration.

Epps, J. A., (1985). "Pavement Type Selection," Texas Hot Mix Association.

Epps, J. A., (1986). "Pavement Thickness Designs for the City of St. Joseph.

Federal Register, (1981). "Pavement Type Selection Policy Statement and Clarification."

Finn, F., Saraf, C., Kulkarni, R., Nair, K., Smith, W. and Abdullah, A., (1977). "The Use of Distress Prediction Subsystems for the Design of Pavement Structures," *Proceedings, Fourth International Conference on Structural Design of Asphalt Concrete Pavements,* University of Michigan, Ann Arbor.

Haas, R., and Hudson, W. R., (1978). *Pavement Management Systems,* McGraw-Hill Book Company.

Lee, R. L., and Grant, E. L. (1965). "Inflation and Highway Economic Studies," *Highway Research Record* No. 100.

Woods, K. B., (1960). *Highway Engineering Handbook,* McGraw-Hill Book Company.

Yoder, E. J., and Witczak, M. W., (1976). *Principles of Pavement Design,* John Wiley and Sons, Inc.

CHAPTER 9

CONSTRUCTION OF LIME STABILIZED BASES AND SUBBASES

9.01 Lime Treatment Methods

Three basic lime treatment processes are recognized: in-place mixing, plant mixing and pressure injection (TRB State of the Art Report No. 5, 1987).

In-place Mixing

In-place mixing may involve: (1) mixing lime with existing material already at the construction site, (2) mixing lime with borrow material off site and then transporting the material to the construction site for final mixing and compaction and (3) mixing lime with the borrow source soil and hauling to the construction site for processing (TRB State of the Art Report No. 5, 1987).

In-place mixing of lime with soil may involve any of the following procedures (TRB State of the Art Report No. 5, 1987):

1. Add a single increment of lime to soils that are easily pulverized and mixed. In this procedure, the mixing and compaction procedure is one operation with no mellowing period required.
2. Add a single increment of lime and allow the mixture to mellow for a period of 1 to 7 days in order to assist in breaking down highly plastic, heavy clays and to facilitate further mixing.
3. Add a single increment of lime as a pre-treatment of the soil prior to treatment with a second stabilizer such as portland cement, fly ash or asphalt.
4. Add a single increment of lime to modify the soil so that the soil will function as a working platform for further construction. Proof rolling is usually required in lieu of more time-consuming and detailed specifications relating to pulverization and density requirements.
5. Add two increments of lime to difficult to pulverize soils with an intermediate period of mellowing between applications to allow breakdown and amelioration of the problem soil.
6. Add lime in a deep lift which may be accomplished in one of two approaches (TRB State of the Art Report No. 5, 1987).
    a. Add one increment of lime to modify the soil to a depth of approximately 610-mm (24-inches). A second increment of lime is added to the top 152 to

304-mm (6 to 12-inches) complete the stabilization process. In this process the break down operation may be accomplished by heavy disc harrows or plows with deep rippers or with new heavy-duty mixing equipment such as that shown in Figure 9.1.

b. Add a single increment of lime for complete stabilization of the soil to a depth of approximately 457-mm (18-inches). Mechanical mixers (Figure 9.1) are now available to pulverize the lime-soil to the full depth by progressive cuts as follows: to a depth of 152-mm (6-inches) on the first pass, 230-mm (9-inches) on the second pass, 304-mm (12-inches) on the third pass, 381-mm (15-inches) on the fourth pass and finally 457-mm (18-inches) on the final pass. The full 381-mm (18-inch) depth is then compacted with a heavy conventional roller or a vibratory roller.

Plant Mixing

Plant mixing involves hauling the soil to a central plant where the soil, lime and water are mixed in a mixing plant designed to intimately and thoroughly mix the constituents in a uniform manner. The mixture is then transported to the construction site for further manipulation. Figure 9.2 is an example of a central mixing plant.

Pressure Injection

In order to control swelling of unstable soils in roadbeds or under foundations or as a part of foundation systems, lime slurry has been injected under pressure to depths of 2.15 to 3.08-m (7 to 10-feet). The injection application is usually applied at 1.54-m (5-foot) spacings and attempts are made to place horizontal seams of lime slurry at 203 to 305-mm (8 to 12-inch) intervals. The top 152 to 304-mm (6 to 12-inches) is then usually stabilized by conventional methods.

Petry et al. (1980) utilized a statistically designed field plot to monitor changes in 20 soil parameters before and after single and double-stage lime slurry pressure injections to an effective depth of 2.15-m (7-feet). Petry et al. (1980) found that at the reduction in swell potential and swell pressure was statistically significant at the 5-percent level after both single and double-injection methods.

Pressure injection has been applied successfully to roadbeds, foundations and railroad roadbeds (Thompson (1975), Blacklock (1978)). Details of lime slurry injection are discussed by J. R. Blacklock et al. (1978, 1982, 1977, and 1986) and in the National Lime Association's Bulletin 331 addressing "Lime Slurry Pressure Injection (LSPI)." The concept of LSPI is discussed in this manual in Chapter 12.

CONSTRUCTION OF LIME STABILIZED BASES AND SUBBASES | 163

FIGURE 9.1. EXAMPLE OF MODERN HEAVY DUTY MIXING EQUIPMENT WHICH CAN BREAK DOWN AND PULVERIZE TO 457-MM (18-INCHES) IN A SINGLE PASS.

FIGURE 9.2. EXAMPLE OF A CENTRAL MIXING PLANT IN WHICH HYDRATED LIME IS INTIMATELY MIXED WITH SOIL TO REDUCE THE PI.

9.02 Steps in the Construction Process

Soil Preparation

The subgrade soil must be brought to final grade and alignment. However, the finished grade elevation may require adjustment because of the potential fluff action of the lime-stabilized layer resulting from the fact that some soils tend to increase in volume when mixed with lime and water. This volume change may be exaggerated when the soil-lime mixture is remixed over a long period of time, especially at moisture contents less than optimum moisture for compaction. The fluff action is usually minimized if adequate water is added and mixing is accomplished shortly after lime is added. For soils that tend to fluff with lime, the subgrade elevation should be lowered slightly or the excess material trimmed. Trimming can usually be accomplished by blading the material onto the shoulder of the embankment slopes.

It is important to understand that the volume change or "fluff" in lime-treated soils is due to the significant flocculation effect of lime treatment and not to swell through the hydration of clay layers. The change from a dispersed to flocculated structure is a good change as it reduces swell potential and further volume changes and improves shear strength.

If lime is added in a dry form, ripping and scarification of the natural soil to be stabilized may be accomplished either before or after adding the lime. If the lime is to be added in the form of a slurry, it is better to scarify the natural soil prior to addition of the lime.

Blading may be needed in order to remove the top 6.4-mm (0.25-inch) of lime-soil mixture before placing the pavement layer. This exposed surface layer is subjected to carbonation and loss of lime due to rain and surface water. Therefore, the top 6.4-mm (0.25-inch) or so is often not well cemented (TRB State of the Art Report No. 5, 1987).

Lime Application

Dry Hydrated Lime

Dry hydrated lime can be applied either in the bulk form or by bags. Application of lime by bags is the simplest method but is usually the most costly. This is primarily because the 22.7 Kg (50-pound) bags of lime require a considerable amount of time to hand place and open. After the bags are placed, they are slit open and the lime is dumped into piles or transverse windrows across the roadway. The lime is then either leveled by hand or by a spike-tooth harrow or drag pulled by a tractor or truck. It is necessary to immediately sprinkle the applied lime to reduce dusting (TRB State of the Art Report No. 5, 1987 and NLA Bulletin 326, 1991).

Despite the higher costs and slower operation associated with the use of bagged lime, it is often the most practical method for small projects or for projects in which it is difficult to utilize large equipment.

Details on the use of bag lime in the stabilization operation are presented in the National Lime Association's Bulletin 326—Lime Stabilization Construction Manual.

Bulk lime application is the most common practice for large stabilization projects, particularly when dusting is not a concern. In this operation lime is typically delivered to the job in self-unloading transport trucks (Figure 9.3). These large and efficient trucks are capable of hauling 13.6 to 21.7 metric tons (15 to 24 tons). These trucks may be equipped with either screw conveyors that discharge at the rear or pneumatic units that blow the lime from the tanker compartments through a pipe or hose to a cyclone spreader or to a pipe spreader bar, mounted at the rear (Figure 9.4).

When the auger trucks are used, spreading is handled by means of a portable type spreader attached to the rear of the truck (Figure 9.5) or through downspout chutes extending from the screw conveyors. The mechanical spreaders incorporate belt, screw, rotary vane or drag-chain conveyors to distribute the lime uniformly across the spreader width. When boots or spouts are used, the lime is deposited in windrows: but due to

FIGURE 9.3. LIME TRANSPORT TRUCKS USED FOR THE DELIVERY AND SPREADING OF DRY HYDRATED LIME.

the powdery nature and ease of flow of lime, it becomes distributed rather uniformly across the spreading lane. Regardless of the spreading mechanism selected, the rate of lime application can be regulated by varying the spreader opening, spreader drive speed, or truck speed so that the required amount of lime can be applied in one or more passes (TRB State of the Art Report No. 5, 1987).

Pneumatic trucks generally employ cyclone spreaders mounted to the rear of the truck. This type spreader distributes lime through a split chute or with a spreader bar equipped with several large downspout pipes. Adjustments made to the air pressure through console-mounted controls permit the driver to quickly and efficiently adjust spreading width.

Dry Quicklime

Quicklime, like hydrated lime, can be applied in bags or in bulk. Higher cost bagged lime is only used in drying of isolated wet spots or on small jobs. The distribution of bagged quicklime is similar to that of bagged hydrated lime, except that greater safety

CONSTRUCTION OF LIME STABILIZED BASES AND SUBBASES | 167

FIGURE 9.4. PIPE SPREADER BAR MOUNTED AT THE REAR OF A TANKER USED FOR SPREADING DRY HYDRATED LIME.

FIGURE 9.5. PORTABLE TYPE SPREADER ATTACHED TO THE REAR OF A TRUCK USED FOR SPREADING DRY HYDRATED LIME.

is required in handling the quicklime. With bagged quicklime, emphasis should be placed on quickly watering and mixing of the quicklime with the soil in order to minimize the danger of burns. Quicklime may be applied in the form of pebbles of approximately 9.5-mm (3/8-inch), granular or pulverized.

Bulk quicklime may be spread by self-unloading auger or pneumatic transport trucks, similar to those used for dry hydrate. However, because of its coarser size and higher density, quicklime may also be tailgated from a regular dump truck with tailgate opening controls to ensure accurate distribution.

Quicklime is anhydrous and generates heat on contact with water. Therefore, special care must be taken to avoid burns caused by hydration of the quicklime when water is added. When quicklime is specified, it is important for the contractor to provide the engineer with a detailed safety program covering precautions and emergency treatment available on the job site. The program should include protective equipment for eyes, mouth, nose and skin as well as a first-aid kit with an eyeball wash.

Slurry Method

Slurry Method with Hydrated Lime

Hydrated lime-water slurry is mixed either in a central mixing tank, jet mixer (Figure 9.6) or a tank truck. The slurry is spread over a scarified roadbed by a tank truck equipped with spray bars. One or more passes may be required over a measured area to achieve the specified percentage based on lime solids content.

A typical slurry mix proportion is 0.9 metric ton (1 ton) of lime and 1295 liters (500 gallons) of water, which yields approximately 2271 liters (600 gallons) of slurry containing about 30 percent lime solids. At one time it was difficult to pump and spray concentrations higher than about 31 percent. However, with new equipment and techniques, high-volume hydrated slurry mixes have overcome these problems.

Forty percent solids is a maximum pumpable slurry. The actual proportion used in the slurry depends on the percentage of lime specified, type of soil being stabilized and its moisture condition. When small lime percentages are required, the slurry proportions may be reduced to 0.9 metric ton (1 ton) of lime per 2650 to 3028 liters (700 to 800 gallons) of water. When the moisture content of the soil is near optimum, a stronger lime concentration would normally be required (TRB State of the Art Report No. 5, 1987).

Central mixing plants employ agitation by using compressed air and a recirculating pump. The most typical slurry plant incorporates slurry tanks large enough to handle whole tank truck loads of hydrated lime of approximately 18 metric ton (20 tons).

Compact jet slurry mixers are an efficient method of slurry production. Water at 482 kPa (70 psi) and hydrated lime are charged continuously in a 65:35 weight ratio into the jet mixing bowl where slurry is produced instantaneously. The mixer and auxiliary

FIGURE 9.6. MIXING OF DRY HYDRATED LIME WITH WATER IN A JET MIXER TO PRODUCE A LIME SLURRY.

equipment can be mounted on a small trailer and transported to the job readily, giving great flexibility to the operation (TRB State of the Art Report No. 5, 1987).

A third type of slurry operation mixes measured amounts of water and lime which are separately charged into a tank truck with the slurry being mixed in the tank either by compressed air or by a recirculating pump mounted at the rear. The water is metered and the lime proportioned volumetrically or by means of weight batchers. Both portable and permanent batching plants are used. Mixing with air is accomplished at the plant. The air jets are turned on during the loading operation, and remain on until the slurry is thoroughly mixed which takes about 10 to 15 minutes. The use of recirculating pumps permits mixing to occur during transit to the job (TRB State of the Art Report No. 5, 1987).

Spreading from the slurry distributors is effected by gravity or by pressure spray bars, the latter being preferred because of better distribution. The use of spray deflectors is also recommended for good distribution. The general practice in spreading is to make either one or two passes per load. However, several loads may be needed in order to distribute the required amount of lime. The total number of passes will depend on the lime requirement, optimum moisture or the soil and type of mixing employed. Windrow mixing with the grader generally requires several passes (TRB State of the Art Report No. 5, 1987).

Slurry Method with Quicklime

A slaking unit system has been developed for making lime slurry from quicklime (Figure 9.7). The unit consists of a 3.08-m (10-foot) diameter by 12.3-m (40-foot) tank that incorporates a 1.54-m (5-foot) diameter single shaft agitator turned by a 100-hp diesel engine (McKennon, 1990). The Portabatch slaker can handle 18.1 to 22.7 metric tons (20 to 25 tons) of quicklime and about 94 m^3 (25,000 gallons) of water, producing the slurry in about 1 to 1.5 hours. Because of the exothermic action of quicklime in water, the slurry is produced at a temperature of about 85°C to 93.3°C (185°F to 200°F).

In the quicklime-slaking process, the tank is first filled with about 58 m^3 (15,000 gallons) of water. A full truckload of quicklime is then discharged pneumatically underneath the water level, with the water serving as a wet scrubber for the lime dust. The paddle system stirs the water as the lime is added. As the lime slakes, the exothermic heat is utilized to maintain the slaking temperature of about 93.3°C (200°F). This produces a small hydrated lime particle with high surface area. In this process, the quicklime slakes into hydrated lime with no unhydrated particles remaining. The entire reaction is completed within 10 minutes after the last lime is unloaded from the truck. Thus, a 22.7 metric ton (25 ton) load of quicklime is converted into about 29.5 metric tons (32.5 tons) of hydrated lime on a dry basis, suspended in a slurry with 30 to 35 percent solids, all within about 1 hour.

A serious operating problem originally encountered with quicklime slaking was accurately arriving at a weighing method on which to base the payment of lime. Most specifications express the basis of payment on tons of dry hydrated lime used. In this method the problem occurs because lime is manufactured at the job site where quicklime (CaO) is converted to hydrated lime ($Ca(OH)_2$). Forty-five Kg (100 pounds) of quicklime converts into 58 to 59 Kg (129 to 131 pounds) of hydrated lime depending on the purity of the incoming quicklime or at a 100:129 or 100:131 quicklime to hydrate ratio.

Calculation of converted weight of hydrated lime is based on the Calculated Method. With this procedure, each load of incoming quicklime is checked for available CaO prior to leaving the quicklime producing plant. Using a standardized molecular weight ratio, the exact tonnage of hydrated lime in the slurry can be calculated. A 90 percent CaO tested truckload converts into dry hydrated lime at a 128.8 ratio; a 93 percent available CaO truckload, which is typical of the quicklime produced by most lime manufacturers, converts at a 129.7 ratio, and a 95 percent available CaO truckload converts at a 130.4 ratio. This method is now widely used in many states and has

Figure 9.7. Two Views of Portabatch Slaking Unit Used in the Production of Lime Slurry from dry Quicklime. This Operation Was Used on the Massive Denver International Airport Project in 1991–1993.

produced reliable results. The chemical conversions for quicklime to hydrated lime are given below:

Available CaO, Percent	Pounds of Dry $Ca(OH)_2$ per 100 pounds CaO
90.0	128.8
93.0	129.7
95.0	130.4

1 pound = 0.454 Kg

It is now possible to produce high quality lime slurries at the job site with either a quicklime slaking process as previously discussed or with high-volume slurry mixes which mix bulk hydrated lime to produce high volume slurries.

Advantages and Disadvantages of Different Lime Applications

The selection of the best type of application must be based on the relative advantages and disadvantages of each. Some of these are listed here and are taken directly from the TRB Circular 180, State of the Art Report 5 (1987):

Dry hydrated lime

   a. Advantages:

     Dry lime can be applied two or three times faster than a slurry and

     Dry lime is very effective in drying out wet soils.

   b. Disadvantages:

     Dry lime produces a dusting problem that makes its use undesirable in urban or closely contained areas and

     The fast drying action of dry lime requires an excess amount of water during the dry, hot seasons.

Dry quicklime

   a. Advantages

     Quicklime is more economical as it contains approximately 30 percent more available lime,

     Quicklime has greater bulk density for smaller-sized storage silos,

     Quicklime has faster drying action in wet soils and is faster reacting with soils and

The construction season can be extended in both Spring and Fall because the heat generated upon slaking of quick lime warms the soil,

Quicklime reacts faster with soils.

b. Disadvantages:

Field hydration may be less effective than commercial hydrators, producing a coarser material with poorer distribution in the soil mass, however slaking of quicklime to form a quick lime slurry is a very effective method of hydration,

Quicklime requires more water than hydrate for stabilization which may present a problem in dry areas and

Greater susceptibility to skin and eye burns.

Slurry lime

a. Advantages:

Dust-free application is more desirable from an environmental standpoint,

Better distribution is achieved with slurry,

In lime slurry, the lime spreading and sprinkling operations are combined, thereby reducing costs,

During summer months slurry application pre-wets the soil and minimizes drying action. And requires less final mix water.

b. Disadvantages:

Application rates may be slower,

Extra equipment is required, therefore, costs are higher,

Extra manipulation may be required for drying purposes during cool, wet, humid weather, which could occur during the fall, winter and spring construction season.

May not be practical for use with very wet soils.

Pulverization and Mixing

Adequate pulverization and mixing are absolutely essential to achieve satisfactory results in lime stabilization. While most soils may only require one-stage mixing, heavier, more plastic soils require multiple mixing.

One-Stage Mixing

One stage mixing can be effectively achieved with either blade or rotary mixing or a combination of both. Rotary mixing is preferred as it provides the most uniform mixing, finer pulverization and a faster mixing process.

Two-Stage Mixing

The two-stage process consists of preliminary mixing, moist curing for a period ranging from about 24 to 72 hours or even more and final mixing or remixing. In the preliminary mixing operation, the objective is to distribute the lime throughout the soil and thereby allow the mellowing operation to take place. In order to optimize the chemical reactions of cation exchange and some pozzolanic reactivity necessary to facilitate the mellowing process, it is necessary to break the clay clods down to sizes of less than 51-mm (2-inches) in diameter. Prior to mellowing the soil should be watered liberally to bring it to at least two percentage points above optimum, which aids in the breakdown or disintegration of the clay (TRB State of the Art Report No. 5, 1987).

In hot weather it is difficult to add too much water; however, in wet, cool weather it may be necessary to adjust the water used in preliminary mixing to somewhat lower levels.

As soon as possible following preliminary mixing the roadway should be sealed by lightly rolling with a rubber-tired roller. This procedure provides a compacted subgrade that will shed water and prevent the soil from taking on water which could result in construction delays. Generally after a 24 to 48 hour delay the clay becomes friable enough to easily achieve desired pulverization in final mixing. Additional wetting or sprinkling may be necessary during final mixing to bring the soil to optimum moisture or slightly above. In hot weather more than optimum moisture is needed to compensate for the loss through evaporation.

Disc harrows (Figure 9.8) and grader scarifiers (Figure 9.9) may be suitable for preliminary mixing. However high-speed rotary mixers (Figure 9.10) are required for final mixing. Motor graders are generally unsatisfactory for mixing lime with heavy clays (TRB State of the Art Report No. 5, 1987).

Blade Mixing

When blade mixing is used with dry lime, the soil is generally bladed in two windrows, one on each side of the roadway. Lime is then spread between the two windrows. The soil is next mixed by blading it over the lime. After covering the lime with the soil, mixing is continued by blading across the roadway. After dry mixing is completed, water is added to slightly above the optimum moisture content and additional mixing is performed (NLA Bulletin 326, 1991).

When blade mixing a slurry and soil mixture, the mixing is done in thin lifts that are bladed to windrows. One procedure is to start with the material in a center windrow, then blade aside a thin layer after the addition of each increment of slurry as side windrows are formed. This windrowed material is then bladed back across the roadway

FIGURE 9.8. DISC-HARROWS USED IN INITIAL MIXING OF LIME IN A TWO-STAGE MIXING PROCESS.

FIGURE 9.9. GRADER-SCARIFIERS USED IN INITIAL MIXING OF LIME IN A TWO-STAGE MIXING PROCESS.

FIGURE 9.10. HIGH-SPEED ROTARY MIXERS REQUIRED FOR FINAL MIXING IN A TWO-STAGE MIXING PROCESS.

and compacted, provided that its moisture content is optimum (NLA Bulletin 326, 1991).

An alternate practice is to start with a side windrow and then blade a thin 51- mm (2-inch) layer across the roadway and add an increment of lime, then blade this layer to a windrow on the opposite side of the road (NLA Bulletin 326, 1991).

Rotary Mixing

When high-speed rotary mixers are used, lime is usually spread evenly on the entire roadway, and mixing is begun from the top down. Complete mixing is normally accomplished in one to three passes, as long as adequate equipment is being used, with most soils. The desired mixing moisture content for the rotary mixing operation is usually approximately optimum. The water may be added by sprinkling trucks or by spraying into the mixing chamber of the mixer. The latter method is preferred as it provides more intimate contact of lime, water and soil and facilitates chemical breakdown and pulverization (NLA Bulletin 326, 1991).

Central Mixing and Upgrading Marginal Aggregates with Lime

Upgrading marginal granular base material with lime is becoming popular. The benefits of this process of upgrading the marginal material becomes more evident as the result of research (see Chapter 6).

The gravel or granular base usually must be processed to meet gradation specifications. It is then a relatively simple matter for the contractor to install a lime bin, feeder and pug mill at the screening plant. Figure 9.11 is a schematic of the simple components required for such an operation.

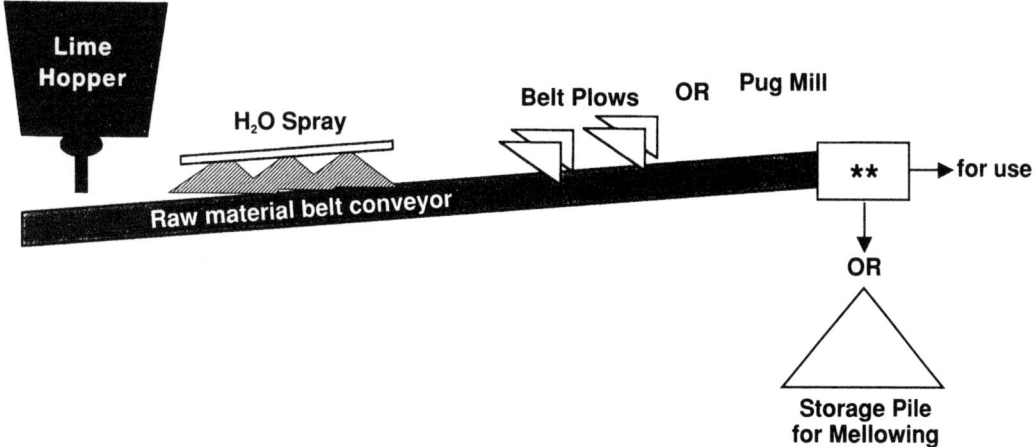

FIGURE 9.11. SCHEMATIC ILLUSTRATING THE SIMPLE BASIC COMPONENTS REQUIRED FOR A CENTRAL MIXING FACILITY FOR MIXING LIME WITH MARGINAL AGGREGATE.

In some applications lime is used to modify aggregate in which the fines content (minus 40 sieve fraction) is relatively high as is the plasticity of the fines as denoted by the plasticity index (PI). Lime is added to lower the PI or to reduce it to a non-plastic state. A second application is to add lime to the aggregate to remove a clay film from the aggregate which is to be used in production of portland cement concrete, asphalt concrete or ballast and sand. Besides reducing the PI and increasing the sand equivalent, the addition of lime to aggregates containing plastic fines can improve mixture response and structural characteristics as discussed in Chapter 6. In addition the production life of certain sand and gravel pits and crushed stone quarries can be extended for years by successfully modifying the plastic fines and producing aggregates with marketable qualities.

For aggregates with PI's of less than 20, lime contents of between 0.5 to 1.5 percent (by weight of the total aggregate) are typically used. For aggregates with PI's of greater than 20, lime requirements of between 1 and 3 percent are typically required.

The following equipment is needed to feed the lime to the aggregate at the processing plant: storage bin for the quicklime, variable speed vane feeder to deposit the lime into the aggregate (and water distribution system to hydrate the quicklime completely if quicklime is added instead of hydrated lime).

If quicklime is added, a water distribution system consisting of about 3.1-m (10-feet) of 12-mm (1/2-inch) pipe is installed. The pipe is drilled on 203-mm (8-inch) centers and tapped to accept high-volume spray nozzles. The distribution pipe should be centered over the conveyor belt, just beyond the lime feeder outlet so it will be in position to spray the quicklime and hydrate the material completely

A series of belt plows is recommended to mix the lime thoroughly with the aggregate. It is also recommended to install a "flipper" or "paddle wheel" device or a pug mill at the final discharge head pulley. This device will improve the lime/aggregate mixing action and prevent segregation.

After the lime has been added, slaked and properly mixed with the aggregate, a curing period of approximately 48 hours is recommended to insure the complete reaction of the quicklime with the clay portion of the treated material. Stockpiling is the normal procedure.

During and beyond the curing period, the lime agglomerates the clay fines into coarse, friable particles. This makes the fine clay particles stick together and become large enough to settle during a sand equivalent (SE) test.

Advantages of the lime-upgraded bases include:

1. Water resistance,
2. Greater strengths,
3. Improved stiffness and structural contribution,
4. Improved durability and
5. Improved consistency and uniformity.

Pulverization and Mixing Requirements

Pulverization and mixing requirements are usually specified in terms of the percentages passing the 38.1 or 25.4-mm (1.5-inch or 1-inch) sieve and the Number 4 sieve. Typical requirements are 100 percent passing the 25.4-mm (1-inch) and 60 percent passing the Number 4 sieve, exclusive of non-slaked fractions. In certain expedient construction operations, formal requirements are eliminated and the "pulverization and mixing to the satisfaction of the engineer" clause is used.

Compaction

To insure maximum development of strength and durability it is important that the mix is properly compacted (See Figure 9.12). Many agencies require at least 95 percent of AASHTO T-99 density for subbases and 98 percent for bases. Some agencies have required 95 percent of AASHTO T-180 maximum density. Although such densities can be achieved for most granular soil-lime mixtures, it is difficult to achieve this degree of compaction for lime-treated fine-grained soils.

Thick lifts of lime-soil mixture may be compacted successfully in a single compaction application. A typical specification for such an operation is 95 percent of AASHTO T-99 maximum density in the upper 152 to 229-mm (6 to 9-inches) and lower densities are accepted in the lower portions of the lift.

For granular soils compaction should begin as soon as possible after mixing, although delays of up to 2 days are not normally detrimental. However, if the soil is allowed to become too dry during the mellowing period and if excessive carbonation is allowed to occur, the results can be deleterious. Fine-grained soils can also be compacted soon after mixing although delays of up to approximately 4 days are not normally detrimental. When longer delays (2 weeks or more) cannot be avoided, it may be necessary to incorporate a small additional amount of lime into the mixer to compensate for lime lost due to dusting and/or carbonation (TRB State of the Art Report No. 5, 1987). When double applications of lime with extended delay or mellowing periods are required to guard against troublesome soils containing high sulfate levels, consult Section 4.07 of this manual.

The most common practice in lime stabilization is to compact the stabilized layer in one lift by first using the sheeps-foot roller (Figures 9.13 and 9.14) until it "walks out" and then to use a multiple-wheel rubber-tired roller (Figure 9.14). In some cases, a flat wheel steel roller is used for finishing. Single lift compaction can also be accomplished with a vibrating impact roller or with heavy rubber-tired rollers and light rubber-tired rollers or steel wheel rollers used for finishing (TRB State of the Art Report No. 5, 1987).

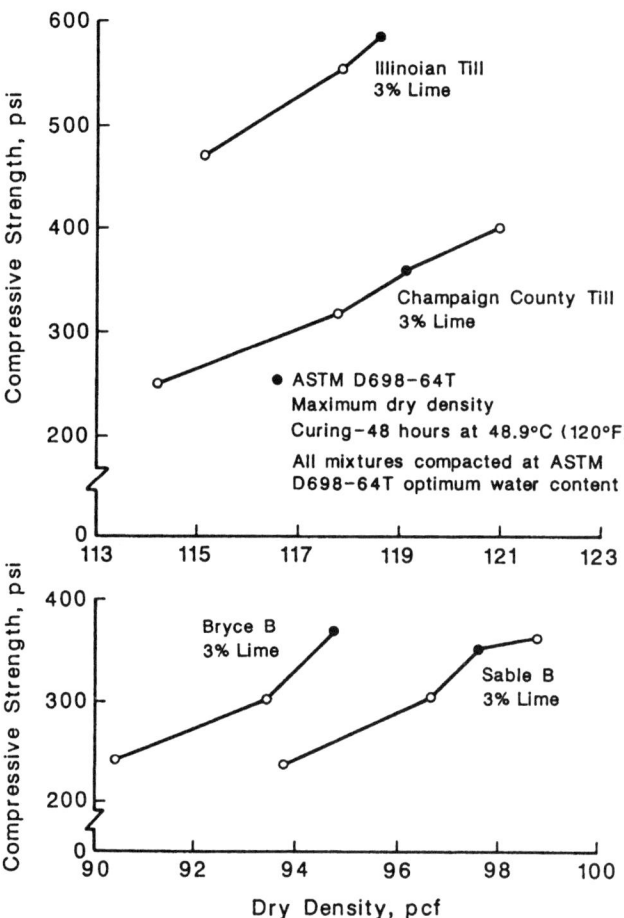

FIGURE 9-12. THE LEVEL OF COMPACTION PRODUCED DURING CONSTRUCTION HAS AN IMPORTANT EFFECT ON IN PLACE STRENGTH OF THE LIME STABILIZED PAVEMENT LAYER (AFTER THOMPSON, 1969). (1 pct = 16 Kg/m^3)

Curing

To achieve maximum strength and durability proper curing as well as good compaction is required. Temperatures higher than 5°C to 15°C (40 to 50°F) and moisture contents around optimum for compaction are necessary for adequate curing. Some agencies require a period of curing of between 3 and 7 days prior to placing upper structural layers. However, other agencies allow immediate application of the upper layers. This approach has the advantage of protecting the lime-soil mixture and inducing curing if the lime-soil mixture is strong enough, at the time of upper layer construction, to resist rutting (NLA Bulletin 326, 1991).

CONSTRUCTION OF LIME STABILIZED BASES AND SUBBASES | 181

Figure 9.13. Sheeps-foot Rollers Are Used to Compact the Lime-Treated Layer in One Lift. Rolling Is Continued Until the Roller "Walks Out".

Figure 9.14. After the Sheeps-Foot Roller "Walks Out" a Multi-Wheeled Pneumatic Roller Is Normally Used to Complete Compaction to Meet Specifications. In Some Cases, Single Lift Compaction Can Be Accomplished With Heavy Pneumatic Rollers With Vibrating Impact Rollers.

FIGURE 9.15. MOIST CURING IS PROVIDED BY KEEPING THE SURFACE DAMP BY SPRINKLING. THIS IS A DEMANDING TASK IN DRY CLIMATES OR UNDER DRY CONDITIONS.

Two types of curing can be used: moist and asphaltic membrane. In the first, the surface is kept damp by sprinkling (Figure 9.15) with light rollers being used to keep the surface knitted together. In membrane curing, the stabilized soil is either (a) sealed with one shot of cutback asphalt at a rate of about 0.10 to 0.25 gallons per square yard within one day after final rolling, or (b) primed with increments of asphalt emulsion applied several times during the curing period. A common practice is to apply two shots asphalt emulsion the first day and one each day thereafter for 4 days at a total rate of 0.10 to 0.25 gallons per square yard (Figure 9.16).

9.03 Construction Considerations for Lime and Fly Ash Applications

Among the advantages of lime—fly ash (LFA) mixtures for use in pavement construction are the ease of construction and the fact that no special construction equipment is needed. The major requirements during construction for the effective use of LFA materials are that they be well mixed, spread uniformly to the proper thickness

FIGURE 9.16. IN MEMBRANE CURING THE STABILIZED SOIL IS EITHER SEALED WITH CUTBACK OR EMULSIFIED ASPHALT AT THE RATE OF APPROXIMATELY 0.10 TO 0.25 GALLONS PER SQUARE YARD WITHIN ONE DAY AFTER FINAL MIXING. SUBSEQUENT APPLICATIONS OF EMULSION ARE OFTEN REQUIRED.

and compacted to a high relative density. These operations can all be accomplished with construction equipment normally found on a pavement construction site. While the accepted construction procedures are fairly simple, it is emphasized that poor construction procedures will result in poor quality in the final product with a concomitant poor reliability in performance.

Successful central plants for the production of LFA mixtures require the following main components:

1. An aggregate source and hopper with a feed control,
2. A hopper and feed belt with controls for the fly ash,
3. A lime storage unit with feed hoppers and feed controls,
4. Water storage tanks with feed control and
5. A pugmill for blending the materials.

There have also been excellent jobs placed in which the ingredients were spread on a prepared roadbed and mixed in place. Experience with the mixed in place type of blending, however, shows that the overall quality of the final mixtures is not as high as when the materials are proportioned and blended in a central plant.

Spreader boxes have been successfully used to spread the delivered LFA materials. An alternate procedure which has performed successfully is to dump the prepared mixture from the truck into windrows and spread with a motor grader. Particular care must be taken with this method of operation to prevent segregation of the aggregate particles by sizes. A third method of spreading the mixture which provides a high degree of thickness control is to place the mixture with equipment which controls the level of the spread mix by a string-line or similar screed elevation control. With all methods, care must be taken to produce a layer of uniform thickness and to prevent segregation of material during dumping and spreading.

Steel wheeled, pneumatic tired and vibratory rollers have all been used successfully to compact LFA mixtures. Vibratory pan type compactors are also effective for this operation. Since the material is basically granular in nature with little or no cohesion, pneumatic and vibratory rollers and the vibratory pans are usually most effective in producing the desired high relative densities. Steel wheeled rollers are normally used for producing a true and smooth final surface after initial compaction with the other types of compactors.

One of the advantages of LFA mixtures over other stabilized materials is that they can be effectively compacted at any time after mixing up to 24 hours or more. Compaction 4 to 8 hours after mixing is quite common, and there have been cases in which compaction was completed more than 48 hours after the material had been mixed and placed. The length of time that can elapse between mixing and final compaction is a function of climatic conditions. Generally, delays of 24 hours or more are to be discouraged.

Obtaining a high degree of relative density is an absolute necessity for obtaining a quality product. Relative density has a profound effect on both compressive strength and durability.

In general, traffic can be permitted on a compacted LFA mixture immediately after placement. To reduce the abrasive effects of traffic, it is recommended that a surface course be placed over LFA material as soon as possible. The surface course will also prevent evaporation of moisture from the surface.

9.04 Measurement and Payment

The amount of lime used in construction is usually measured in tons. The processing of the lime-treated layer is measured in square meters (square yards), and the amount of water used for mixing, compaction, finishing and curing is measured in units of 3.78 m³ (1,000 gallons). Bituminous materials used for curing seals are measured by the ton or by the gallon.

The basis of payment for lime is at the unit bid price per ton accepted in place. The processing of the lime-treated material is paid for at the unit bid price per square yard of material completed in place. Water is paid for at the unit bid price per 0.378 m³ (100 gallons) of material used on the project, and bituminous membrane material is paid for at the unit bid price per ton or per gallon of material used for curing purposes.

9.05 References

Blacklock, J. R., and Lawson, C. H., (1977). "Handbook for Railroad Track Stabilization Using Lime Slurry Pressure Injection," Federal Railroad Administration Report No. FRA/ORD-77/30.

Blacklock, J. R., (1978). "Evaluation of Railroad Lime Slurry Stabilization," Federal Railroad Administration, Final Report, FRA/ORD-78/09.

Blacklock, J. R., Joshi, R. C., and Wright P. J., (1982). "Pressure Injection Grouting of Landfills Using Lime and Fly Ash," *Proceedings, ASCE Specialty Conference, Grouting in Geotechnical Engineering*, New Orleans.

Blacklock, J. R., and Wright P. J., (1986). "Injection Grouting Restoration of Failed Highway Embankment Slopes," 65th Annual Transportation Research Board, Washington, D. C.

National Lime Association. *Lime Slurry Pressure Injection Bulletin*, Bulletin 331.

National Lime Association, (1991). *Lime Stabilization Construction Manual*, Bulletin 326.

Petry, T. M., and Armstrong, J. C., "Relationships and Variations of Properties of Clay," *Proceedings of the Fourth International Conference on Expansive Soils*, ASCE, Denver, Colorado.

Thompson, M. R., (1969). "Engineering Properties of Lime-Soil Mixtures," *Journal of Materials*, ASTM Volume 4, No. 4.

Thompson, M. R., (1975). "Field Evaluation of Pressure Injection Lime Treatment for Strengthening Subgrade Soils," *Proceedings of Roadbed Stabilization Lime Injection Conference*, FRA-OR&d-76-137.

Transportation Research Board, (1987). *State of the Art Report No. 5, Lime Stabilization*.

CHAPTER 10

QUALITY CONTROL AND GUIDE SPECIFICATIONS FOR LIME-TREATED LAYERS

10.01 Field Control Considerations

A number of factors are critically important to the control of the quality of soil-lime mixtures during construction. These factors are discussed in the following paragraphs.

Lime Spread Rate

The spread rate of lime is determined in terms of pounds of lime per unit area of surface. Probably the most direct and simplest way to determine the actual field spreading rate is to place a 1-square yard piece of material on grade and, following lime spreading, measure the actual spreading rate as the weight of lime on the square yard of material for dry lime. For slurry applications, calculate the gallons per unit area spread times the percent lime solids content.

Pulverization

In most specifications, the efficiency of pulverization is determined based on the amount of material passing the 25.4-mm (1-inch) sieve and the Number 4 sieve. The processed material is dry sieved to determine the percent passing each sieve. Care must be taken to insure that the plus Number 4 material fraction is not actually agglomerated soil-lime mixture that can be easily broken down by a simple kneading action to pass the Number 4 sieve.

Mixing Efficiency

The efficiency of field mixing is of critical importance as the basic reactions cannot proceed successfully nor optimally without this efficient mixing. A simple procedure for evaluating mixing efficiency is:
  a. Secure a sample of the field mixed soil-lime material;
  b. Halve the sample;
  c. Prepare strength specimens for unconfined compressive strength testing (field mix sample) from one portion;
  d. Completely remix the other portion of the field mixture to insure almost 100 percent mixing;

e. Prepare strength specimens from the remixed material;
f. Cure both sets of strength specimens and test them; and
g. Calculate the mixing efficiency as

$$\text{Percent Mixing Efficiency} = \frac{\text{field mixed strength}}{\text{laboratory mixed strength}} \times 100$$

For mixed in-place operations, mixing efficiencies normally range from 60 to 80 percent. In some types of soil-lime mixing operations lower values may be acceptable.

Depth of Lime Treatment

Phenolphthalein is a color-sensitive indicator of pH. Since soil-lime mixtures demonstrates an elevated pH, the indicator can be successfully used to indicate the presence of lime. The indicator is used by spraying it on the soil-lime mixture. If lime is present a reddish-pink color develops.

Lime Content

An ASTM procedure (ASTM D3155-73) has been developed for determining the lime content of uncured soil-lime mixtures. The procedure is rapid and easy to conduct. Other methods of determining lime content are also used.

Moisture Content

Moisture content is determined by using conventional procedures such as oven drying and nuclear methods. When the nuclear density gage is used, it is important to insure calibration for the soil-lime mixture.

Density

Conventional procedures, such as the sand cone, rubber balloon, and nuclear density gage, are used to determine in-situ density of compacted soil-lime mixtures. It is important to insure that the proper moisture-density relation for the soil-lime mixture is used in density control. The moisture-density relation for soil-lime mixtures may change with curing time and variation in lime content. An example of this is when the soil-lime mixture is reworked at some date following initial construction. The maximum dry density and optimum moisture content for the mixture will probably be different from the original mixture.

Slurry Composition

The most convenient method for determining the lime slurry composition is to determine the specific gravity of the slurry by using either a hydrometer or a volumetric-weight procedure. It is important to know the actual quantity of lime slurry required to provide a desired amount of lime solids.

Weather Limitations

The subgrade should not be constructed when weather conditions detrimentally affect the quality of the materials. Lime should not be applied unless the air temperature is at least 5°C (40°F) in the shade and rising. Lime should not be applied to soils that are frozen or contain frost. If the air temperature falls below 2°C (35°F) in the shade, protect completed lime-treated areas by approved methods against the detrimental effects of freezing. Remove and replace or recompact, as indicated, any damaged portion of the completed lime-treated area in accordance with this specification at no additional cost to the owner.

10.02 Guide Specifications for Lime-Treated Subgrades

1.0. General

1.1 General Requirements:

The subgrade soil shall be scarified and mixed uniformly with lime and water, pulverized, shaped, compacted and cured in accordance with these specifications and in conformity to the lines, grades and dimensions as shown on the engineering plans.

1.1.1 Lime Requirement: The percent of hydrated lime/quicklime by weight of dry soil material shall be ____ percent, based upon test results as subsequently outlined.

1.2 Submittals:

1.2.1 Samples: A sample of lime to be used on the job is to be submitted to the Engineer for approval at least seven days prior to the start of lime construction work. Samples shall be submitted in moisture-proof, airtight containers.

1.2.2 Certified Test Reports:

a. Maximum Density and Optimum Moisture ASTM D1157 (D2922) (D3017) with and without recommended lime dosage.

b. Final Compaction Test Reports.

1.2.3 Equipment Lists: Ten days prior to the commencement of the work, submit for approval a list of the equipment to be used and their relationship to the method of mixing, proportioning, application, pulverizing and compacting the subgrade and all other work.

1.2.4 Manufacturer's Certificate of Conformance:

a. Bituminous curing seal

b. Lime, including purity certification

c. Material Safety Data Sheet

1.3 Delivery and Storage:

Deliver lime and bituminous materials in containers showing or including designated trade name, product identification, specification number, manufacturer's name, and source. Store in a manner that will prevent moisture damage, overexposure and contamination.

1.4 Weather Limitations:

Do not construct subgrade when weather conditions detrimentally affect the quality of the materials. Do not apply lime unless the air temperature is at least 5°C (40°F) in the shade and rising. Do not apply lime to soils that are frozen or contain frost. If the air temperature falls below 3°C (35°F) in the shade, protect completed lime-treated areas by approved methods against the detrimental effects of freezing. Remove and replace or recompact, as indicated, any damaged portion of the completed soil-lime treated area, at no additional cost to the owner.

2.0 MATERIALS

2.1 Lime:

Lime shall conform to the following requirements:

2.1.1 Quicklime

2.1.1.1 Quicklime shall contain a minimum of 90 percent calcium and magnesium oxides (on an LOI-free basis) in compliance with ASTM C977. The use of dolomitic quicklime or magnesia quicklimes with magnesium oxide content in excess of four percent are not permitted. For high calcium quicklimes, the minimum available lime index (ASTM C25) shall be 90 percent available CaO. When dry sieved in a mechanical sieve shaker for ten minutes, +/-30 seconds, a 250 gram test sample of quicklime shall conform to the following grading requirements:

Sieve Size	Percent Passing
1-inch (25.4-mm)	100
3/4-inch (19-mm)	95–100
No. 100	0–25

2.1.1.2 High calcium quicklime slurry shall be manufactured from specified quicklime in accordance with 2.1.1.1 except the gradation requirements are waived and subject to approval by the Engineer or Specifier. Dolomitic or magnesia quicklimes shall not be permitted to produce quicklime slurry.

2.1.1.3 When high calcium quicklime (2.1.1.1) is used to produce lime slurry, the quicklime shall be slaked with water to a percent solids acceptable to the engineer (generally between 25-40% solids) and transported to the job site in clean trucks. Lime slurry may be prepared at the job site by slaking high calcium quicklime with water to a percent solids acceptable to the Engineer (generally between 25-40%).

2.1.1.4 The pay basis for dry quicklime supplied directly to the ground shall be per ton of dry quicklime delivered to the job site. The pay basis for hydrated lime slurry, produced from quicklime at the job site, shall be the dry tons of hydrated lime produced at the job site, calculated by an approved, recognized method.

2.1.2.1 High calcium hydrated lime shall contain a maximum magnesium content, calculated as magnesium oxide, of four percent by weight, comply with the chemical composition requirements of ASTM C977, and have an available lime index (ASTM C25) of a minimum 68 percent available CaO (quicklime is 90 percent CaO, and hydrated lime is 68 percent CaO).

2.1.2.2 Dolomitic hydrated lime shall contain a maximum magnesium hydroxide content of 41 percent by weight, and comply with ASTM C977, total unhydrated oxide content (ASTM C25), CaO + MgO, shall be less than five percent, and have an available lime index of 68 percent available CaO.

2.1.2.3 Hydrated lime slurry shall, on a dry basis, conform to the chemical and physical requirements of ASTM C977.

2.1.2.4 The pay basis for slurry produced from dry hydrated lime shall be per ton of dry hydrated lime delivered to the job site.

2.1.2.5 The use of waste hydrated lime, lime kiln dust, or cement kiln dust shall not be permitted.

2.2 Soil:

Soil shall consist of the natural materials in the area to be stabilized (unless otherwise indicated). Remove stones retained on a three-inch sieve and deleterious substances such as sticks, debris, and vegetable matter.

2.3 Water:

Potable or subject to approval by the Engineer. Water high in Sulphates are not suitable for use.

2.4 Bituminous Curing Seal:

    2.4.1 Emulsified Asphalt: Conform to ASTM D977 for Type SS-1 or Type SS-1h; ASTM D2397 for Type CSS-1 or Type CSS-1H. The base asphalt used to manufacture the emulsion shall show a negative spot when tested in accordance with AASHTO T102 using standard naphtha.

    2.4.2 Optional Water Curing: The Contractor upon approval of the Engineer may cure the compacted lime-treated soil layer by keeping the surface *continuously* moist until subsequent layers are applied.

3.0 EXECUTION

3.1 Sequence of Construction Operations

NLA Bulletin 326 (1991), "Lime Stabilization Construction Manual," and TRB (1987), "State of the Art Report 5, Lime Stabilization."

3.2 Site Preparation:

Clean debris from the area to be stabilized. Perform clearing and grubbing (to a specified depth) and recompact as required. Inspect the original ground for adequacy for the forthcoming compactive effort of lime treatment work. Rough grade and shape the area to be stabilized to conform to the lines, grades, and cross sections indicated. (Prior to independent placement of lime-treated course, comply with subgrade requirements of the applicable specifications).

    3.2.1 Grade Control: When the stabilized course is to be constructed to meet with fixed grade, provide adequate line and grade stakes for control. Finished and completed stabilized areas shall conform to the lines, grades, cross section and dimensions indicated. Locate grade stakes in lanes parallel to the center line of areas under construction, and suitably placed for string lining. Maintain the line and grade until further construction prohibits. Contractor should be aware that lime stabilization generally lowers soil density and increases soil thickness and this "fluffing effect" needs to be considered by the contractor to achieve specified final grade sections.

3.3 Lime Treatment:

Comply with NLA Bulletin 326 (1991), "Lime Stabilization Construction Manual," and TRB (1987), "State of the Art Report 5, Lime Stabilization," unless specified otherwise hereinafter.

    3.3.1 General Requirements of Execution: after site preparation, scarify the subgrade and spread lime as specified hereinafter. Blend lime into the subgrade to the required depth as indicated. Apply lime and water to the soil only to those areas where mixing operations can be completed during the same working day. The application and mixing of lime shall be accomplished by either the dry placing method; the slurry method; or by excavation, lime treatment and replacement. No traffic shall be allowed to pass over the spread lime until after completion of mixing. Pulverize, mix, compact, and cure as specified herein. If double application of lime is required because of heavy plasticity of the soil, the percentage of lime for the initial application shall be approximately one-half the total specified lime.

    3.3.2 Scarification: After obtaining required line and grade, scarify and partially pulverize the subgrade. Remove unified organic materials such as stumps and roots, Remove rocks larger than 7,612-mm (3-inches).

    3.3.3 Dry Quicklime Placing: Spread and distribute the quicklime at a uniform rate, with protection from the wind, to the loosely scarified subgrade. The lime shall be applied to the ground in a manner that is uniform with normal construction practice, and shall be subject to approval of the Engineer. After lime is applied to the ground, prevent dry lime from blowing by adding water to the lime or by other suitable means. Do not apply lime when wind conditions, in the opinion of the Engineer, are objectable.

    3.3.4 Dry Hydrated Lime Placing: Dry hydrated lime shall not be applied directly to the subgrade except in remote areas and shall be subject to approval of the Engineer. Do not apply in windy conditions.

    3.3.5 Hydrated Lime Slurry Method: Prepare slurry in a central mixing tank provided with agitation for mixing, if required, to keep the slurry in suspension until applied to the soil. If a slurry jet valve is used for slurring, agitation may be required in central holding tanks and distributor trucks. Spread lime slurry evenly to yield uniform distribution of lime throughout the soil area to be treated. Standard water or asphalt trucks, properly cleaned, with or without pressure distributors, may be used to apply lime treatment.

        The distribution of lime slurry shall be attained by successive passes over the subgrade materials until the proper amount of lime has been spread. The distribution truck shall continually agitate the slurry to keep the mixture uniform or the Contractor shall furnish evidence that his slurry will stay in suspension without agitation. Generally, slurry made directly from dry hydrated lime requires agitation and slurry made by slaking quicklime in special tanks at the jobsite may not require agitation.

After initial mixing and watering, shape and roll the subgrade lightly to seal the surface in order to reduce evaporation of moisture and lime carbonation.

3.3.6 Independent Excavation and Placement of Lime (non-scarification method): Excavate the entire area to receive lime-treated subgrade to the indicated depth. Apply lime; mix; add water; and blend the lime into the stockpiled/imported soil in accordance with the specified and approved blending method. Replace to required indicated grades and compact to at least 95 percent of maximum density at optimum moisture (with lime added) in accordance with ASTM D1557 (AASHTO T180).

3.3.7 Mixing, Uniformity Testing, and Compaction: distribute the lime uniformly in the soil by mixing and pulverizing the subgrade. the rate of spread per linear meter (foot) shall not vary over ten percent from the designated rate. During the mixing process, add water to the subgrade to provide a moisture content of four to five percent above the optimum moisture content and to insure chemical reaction of the lime and subgrade materials (even lime slurry requires additional water). The mixer shall continue making passes until it has produced a homogeneous, uniform mixture of lime, soil and water. After mixing, all the soil particles shall pass a 25.4- mm (one-inch) sieve with at least 60 percent passing the No. 4 sieve. Additionally, the soil-lime mixture shall be free of streaks or pockets of lime and pass the uniformity and gradation test. If dry or slurry hydrated lime is used, moisture is to be at approximately two percent over optimum for all materials other than rock. If dry quicklime is used, moisture shall be at least three percent above optimum for at least 36 hours after application in order to insure that all quicklime is satisfactorily slaked. Compact the lime-treated material immediately after final mixing and testing. Aerate or sprinkle water as necessary to provide adequate moisture conditioning during compaction. Compact the lime-treated material in specified lifts to 95 percent of maximum density at 2 to 3 percent above optimum moisture content in accordance with ASTM D1557 (AASHTO T180). Field density tests shall be based on moisture-density relations of a representative lime-treated sample obtained from the site. As compaction progresses, maintain the shape of the lifts by blading. The surface upon completion shall be smooth and conform to the indicated section and established lines and grades. Perform initial compaction with sheepsfoot or pneumatic rollers. Areas inaccessible to rollers shall be compacted by other means satisfactory to the Engineer.

The entire mixing compaction operation shall be completed within 120 hours of the initial lime spreading, unless otherwise permitted by the Engineer.

3.3.8 Final Curing: If lime slurry is used, no curing period is required beyond the time when gradation is achieved. If dry quicklime is used, cure the lime-treated material for 36-48 hours to be certain that slaking has been completed. During the curing period add water/bituminous curing seal to the surface to maintain the moisture content of the soil-lime mixture at four to five percent above the optimum water content. The soil-lime mixture that has been overexposed to the open air shall be removed and disposed of off-station prior to bituminous sealing or subsequent overlays.

3.3.9 Finishing: The surface of the finished lime-treated soil after compaction shall be the grading plane established, and at any point the surface shall not vary more than 0.05 feet (15.24 mm) above or below the established grade. Any excess lime-treated soil is to be removed. The completed section shall be finished by rolling with a pneumatic or suitable roller sufficiently light to prevent hairline cracking. Keep the surface of each compacted layer of lime-treated soil moist until covered by a subsequent layer of lime-treated material or curing seal.

3.3.9.1 Moist Curing (Water Only): It is mandatory to keep surface damp by sprinkling and use light rollers to keep surface knitted together (preventing surface cracks) until the following course of material is placed.

3.3.9.2 Asphalt Emulsion Curing Seal: Apply at least two applications uniformly to the top (final) layer of lime-treated soil at a rate of 0.15 to 0.20 gallons per square yard of surface. Apply the curing seal as soon as possible after the completion of the final rolling (during the same day), and before the temperature falls below 5°C (40°F).

3.4 Traffic Control, Curing, Maintenance, and Drainage Protection:

Keep all traffic off surfaces freshly treated with bituminous material. provide sufficient warning signals and barricades so that traffic shall not be permitted on the lime-treated soil until stability of the subgrade is assured. Maintain the finished surface until all work has been completed. Provide adequate drainage during the entire period of construction to prevent water from collecting or standing on the area stabilized.

3.5 Equipment Limitations:

3.5.1 General: The Contractor shall submit a list of all construction equipment to the engineer for approval ten days prior to bringing equipment on the job. The type of equipment to be used for each category of work shall conform to the NLA Bulletin 326 or TRB Report 5 unless specified otherwise hereinafter and shall be properly maintained in satisfactory and safe operating condition at all times.

3.5.2 Spreading Equipment: When spreading dry lime at windy locations, use an approved screw-type spreader box, mixer or other semi-enclosed equipment which will offer protection from the wind. Spreading any dry hydrated lime by aggregate spreaders, dump trucks or agricultural spreaders is not allowed except in remote areas with special permission from the Engineer. Field application rate tests may be required at the option of the Engineer who may require the Contractor to change or alter the equipment to be used in the event of non-uniform spreading of lime.

3.5.3 Additional Mixing Equipment Limitations:
  a. Motor graders will not be allowed to mix lime into clays; but, with permission of the Engineer, they may be used for spreading dry quicklime in a uniform manner.
  b. Deep lift rotary mixers may be used, and may facilitate changes in the specified depths of operation, provided the equipment and method of operation sustain a uniform mixture of lime and soil, all to the satisfaction and approval of the Engineer.

3.6 Safety Requirements:

In addition to the Safety Requirements contained within the General Provisions: prevent employee eye or skin contact of quicklime during transport or application. Provide employee use of the following:
  a. Protective clothing, high top boots, gauntlet-type gloves, and protective headgear;
  b. Splash-proof safety goggles and face shields;
  c. Protective cream.

3.7 General:

All sampling and tests shall be performed by an approved testing laboratory at the Contractor's expense (or optionally at owner's expense per contract agreement).

Frequency of sampling and testing of materials for conformance and quality control shall be as specified herein and shall be performed at such other times as necessary to document contract compliance. All test reports and results shall be certified by appropriate test methods.

3.7.1 Optimum Moisture, Maximum Density: The optimum moisture, maximum density test shall be performed on representative samples of the lime-treated soil sampled after final mixing and prior to initial compaction. The soil mixture shall be laboratory compacted within three hours of sampling. Engineer may permit lower compaction results in lime-modified soils or in very deep sections.

3.7.2 Uniformity Tests: After placement and mixing of each lift of lime-treated soil, the Contractor, in the presence of the Engineer, shall perform a series of uniformity tests. The Contractor shall excavate a hole 254-mm (ten inches) in diameter through the full depth of the lift and impregnate the sides of the hole with a standard phenolphthalein indicator. Non-conformity of color reaction will be considered as evidence of inadequate mixing.

3.7.3 Compaction; In-place density to determine the degree of compaction shall be performed between 24 and 48 hours after final compaction. Testing shall be in accordance with ASTM D2922. In-place moisture content may be determined by nuclear method (ASTM D3017) only if the nuclear gauge moisture calibration curve has been established with known moisture contents of the soil to be tested (Sec. 7.1, ASTM D3017).

3.7.4 Thickness and Smoothness: The thickness of the final lime-treated soil shall not vary more than one-tenth foot (1/10) from the design thickness at any point. Final grade smoothness shall not deviate by more than three-eighths inch (3/8) when tested with a ten-foot straightedge.

10.03 Guide Specifications for Lime-Treated Base Courses

The following base stabilization procedure applies to in-place mixing projects:

1.0 Scarification and Pulverization

Same general procedure, as given in 3.2 and 3.3 above, is applicable for in-place base stabilization. In reconstruction of worn-out asphalt roads, the asphalt surfacing should be scarified, broken to pass a two-inch sieve, and mixed with the existing base material. In new construction using transported base material, this step is not applicable.

1.1.1 Equipment: Same as 1.2.3, except a tractor-ripper or "rooter" may be required to break up the old surface.

2.0 Lime Spreading

2.1.1 Scarification: After obtaining required line and grade, scarify and partially pulverize the base material. Remove unified organic materials such as stumps and roots. Remove rocks larger than three inches.

2.1.2 Dry Quicklime Placing: Spread and distribute the quicklime at a uniform rate with protection from wind. After lime is applied to the base, prevent dry lime from blowing by adding water to the lime or by other suitable means. Do not apply lime when wind conditions, in the opinion of the

Engineer, are objectionable. The lime shall be applied to the base in a manner that is uniform within normal construction practice, and shall be subject to approval of the Engineer.

2.1.3 Dry Hydrated Lime Placing: Dry hydrated lime shall not be applied directly to the base except in remote areas and shall be subject to approval of the Engineer. Do not apply in windy conditions.

2.1.4 Hydrated Lime Slurry Method: Prepare slurry in a central mixing tank provided with agitation for mixing, if required, to keep the slurry in suspension until applied to the base. If a slurry jet value is used for slurry, agitation may be required in central holding tanks and distributor trucks. Spread lime slurry evenly to yield uniform distribution of lime throughout the base area to be treated. Standard water or asphalt trucks, properly cleaned, with or without pressure distributors, may be used to apply lime treatment.

The distribution of lime slurry shall be attained by successive passes over the base materials until the proper amount of lime has been spread. The distribution truck shall continually agitate the slurry to keep the mixture uniform, or the Contractor shall furnish evidence that his slurry will stay in suspension without agitation. Generally, slurry made directly from dry hydrated lime requires agitation and slurry made by slacking quicklime in special tanks at the job site may not require agitation.

After initial mixing and watering, shape and roll the base lightly to seal the surface in order to reduce evaporation of moisture and lime carbonation.

2.1.5 Equipment: Ten (10) days prior to the commencement of the work, submit for approval a list of the equipment to be used and their relation to the method of mixing, proportioning, applying, pulverizing, compacting the base, and all other work.

3.0 Mixing and Watering

The mixing and pulverizing, of base materials occurs during this step. Water should only be added (and maintained) up to the optimum moisture content. for quicklime additions, water requirements must also include water for slaking.

As base materials are quite coarse, mixing requirements are for uniformity of the base-lime mixture. It is desirable to remove all plus-two inch aggregate and asphaltic lumps from the base course.

3.1.1 Equipment: If transported base material is used or if the soil binder pulverizes readily, then mixing can be accomplished with a motor grader by windrowing, providing at least three mixing passes are made. Generally, however, rotary mixing is preferable, as in subgrade stabilization.

4.0 Compaction

The same general procedures of 3.3.7 apply except that compaction should be at least 98 percent of the maximum density obtained in the AASHTO T180 test. Light water sprinkling may be necessary to maintain moisture at optimum moisture content.

During final compaction, the base should be shaped to conform with the required lines and grade, as shown on typical sections, and all irregularities should be leveled to form a smooth, dense surface.

    4.1.1 Equipment: Ten (10) days prior to the commencement of work, submit for approval a list of the equipment to be used and its relation to the method of mixing, proportioning, applying, pulverizing, compacting the base, and all other work.

5.0 Curing

The compacted base course should be cured five to seven days, using either moist or asphalt membrane curing, as described above. During curing, heavy traffic loads should be discouraged; but, if allowed, rut marks should be repaired to a smooth condition by rolling as necessary. Before applying a bituminous wearing course, the base should be broomed clean and dampened. This is not essential with portland cement concrete pavement.

10.04 Central Mixed Stabilization Procedures

The following base stabilization procedures apply to central mix projects:

1.0 Central Mixing

The lime and aggregate (and fly-ash, if required), should be fed at uniform, predetermined rates into a pug mill-type mixer. Add water up to the optimum moisture content and mix the materials thoroughly. The central mixing plant facilities should be approved by the Engineer, and periodic checks should be made on the accuracy of feed proportioning.

2.0 Placing Material

The complete base course material should be spread uniformly over the entire applicable surface to the specified thickness with an approved aggregate spreader box. Tailgate dumping and spreading with a motor grader or bulldozer is not recommended.

3.0 Compaction

Same compaction requirements and equipment as above apply.

4.0 Curing

Same curing requirements as above apply.

10.05 Lime Modification

Generally, this category involves, central mixing in the case of granular base materials and in-place mixing in the case of fine-grained subgrade materials.

1.0 Base Material

With in-place mixing, the five steps under Base Specifications apply. Where central mixing is used, the same steps as outlined under Base Specifications apply, namely placing the lime-aggregate mixture with a spreader box, compaction, and curing. the curing period for lime-modified bases may be shortened to two days, since it is not paramount that the treated layer develop as great a strength as with conventional base stabilization.

2.0 Subgrade Materials

The same construction steps as above apply here, except that compaction can follow immediately after mixing. The elimination of the initial curing step is possible because the fine degree of pulverization, as required in conventional stabilization of heavy clays, is to require 100 percent passing a 51-mm (two-inch) sieve, but finer gradation is always a good idea if economically feasible. Disc harrows alone may be adequate for mixing, although rotary mixers are still preferred for the heavier clay soils. Compaction requirements may also be lowered below the 95 percent density, subject to the Engineer's approval. This is particularly true where lime modification is employed for producing a working table; in this case, proof rolling may be all that is required. However, a working table may also be a semi-compacted layer that will support construction equipment. In this case, specify minimum weight-to-support and conditions under which undercut and replacement with fresh lime-treated soil will be added.

Where lime is used to condition a heavy clay soil for stabilization with cement or asphalt, a general procedure is to mix the lime and soil; seal the layer; cure for 24 - 48 hours; remix; then apply the second additive; remix; compact; and cure for seven days.

10.06 Applicable Publications to Lime Stabilization of Subgrades, Subbases or Bases

AMERICAN ASSOCIATION OF STATE HIGHWAY AND TRANSPORTATION OFFICIALS (AASHTO) PUBLICATIONS

M216-84	Quicklime and Hydrated Lime for Soil Stabilization
T87-86	Dry Preparation of Distribute Soil and Soil Aggregate Samples for Test.
T89-86	Determining the Liquid Limit of Soil
T90-86	Determining the Plastic Limit and Plasticity Index of Soil
T102-83	Spot Test of Asphaltic Materials

T193-81(86)   The California Bearing Ratio
T219-86   Testing Lime for Chemical Constituents and Particle Sizes
T220-66(86)   Determination of the Strength of Soil-Lime Mixtures

AMERICAN SOCIETY FOR TESTING AND MATERIALS (ASTM) PUBLICATION, LATEST EDITION

C25	Chemical Analysis of Limestone, Quicklime and Hydrated Lime
C51	Definition of Terms Relating to Lime and Limestone
C110	Physical Testing of Quicklime, Hydrated Lime and Limestone
C977	Quicklime and Hydrated Lime for Soil Stabilization
D558	Moisture-Density Relations of Soil-Cement Mixtures
D559	Wetting and Drying Tests of Compacted Soil-Cement Mixtures
D560	Freezing and Thawing Tests of Compacted Soil-Cement Mixtures
D977	Emulsified Asphalt
D1556	Density of Soil in Place by the Sand-Cone Method
D1557	Moisture-Density Relations of Soils and Soil Aggregate Mixtures Using 10 lb. (4.5 kg) Rammer and 18-inch (457 mm) Drop
D1633	Compressive Strength of Molded Soil-Cement Cylinders
D2397	Cationic Emulsified asphalt
D2487	Standard Method for Classification of Soils for Engineering Purposes
D2922	Density of Soil and Soil-Aggregate in Place by Nuclear methods (Shallow Depth)
D3017	Moisture Content of Soil-aggregate in Place by Nuclear Methods (Shallow-Depth)
D3155	Lime Content of Uncured Soil-Lime Mixtures
D3551	Standard method for Laboratory Preparation of Soil-Lime Mixtures Using a Mechanical Mixer
D3668	Standard Test Method for Bearing Ratio of Laboratory Compacted Soil-Lime Mixtures
D3877	Standard Test Methods for One-Dimensional Expansion, Shrinkage and Uplift Pressure of Soil-Lime Mixtures
D4318	Liquid Limit, Plastic Limit and Plasticity Index of Soils
D5102	Unconfined Compressive of Soil-Lime Mixtures

OTHER IMPORTANT SPECIFICATIONS AND PUBLICATIONS

AASHTO—Interim Specifications for Highway Construction, 1985 (Section 307 on Lime-treated Subgrade).

AASHTO—Interim Specification for Lime for soil Stabilization (M 216-841) (ASTM Designation C977-89).

U.S. Corps of Engineers. "Engineering and Design Manual-Soil Stabilization of Roads and Streets. CE 807.32, December, 1961 (partly revised in February, 1971).

National Lime Association. "Lime Stabilization Construction." Bulletin 326, 1991.

Air Force Engineering and Service Center (AFESC). "Soil Stabilization for Roadways and Airfields," July, 1987.

Transportation Research Board (TRB) "State of the Art Report 5, Lime Stabilization," 1987.

State specifications for the following states: Alabama, Arizona, Arkansas, California, Colorado, Florida, Georgia, Idaho, Illinois, Iowa, Kansas, Maryland, Minnesota, Mississippi, Missouri, Nebraska, New Mexico, New York, North Carolina, North Dakota, Ohio, Oklahoma, South Dakota, Tennessee, Texas, Utah, Virginia, Wisconsin, Wyoming and others.

10.07 Precautions

All engineered construction systems require proper design and proper construction procedures for reliable performance. Lime stabilization has been used in the United States for over fifty years and has an enviable record of successful mitigation of various soil and aggregate problems. This long-term experience has aided in developing standardized procedures designed to minimize design and construction problems.

Some of the more important areas where careful design and field construction practice should be emphasized are as follows:

Amount of Lime

The amount of lime is generally determined by conducting traditional laboratory strength tests using various amounts of lime to establish the optimum amount for each soil. Generalizations and guesses as to the amount of lime to be specified should never be allowed. If insufficient lime is used, soil modification, rather than soil stabilization may occur, reducing the total benefit. The added strength resulting from proper soil stabilization generally is important in roadway design, and careful field monitoring to insure meeting design criteria is important. One method to determine the amount of lime required to achieve soil stabilization is the Eades-Grim pH test, now adopted as an appendix to ASTM C977. This test determines the amount of lime required to achieve a saturated lime solution pH (12.45 at 25°C) in the soil-lime water system after a specified time period. This insures that adequate lime is available to drive pozzolanic reactions needed to develop cementation products in the soil-lime water system. Generally, experienced engineers specify 0.5-1.0 percent more lime than indicated by ASTM C977 to allow for minor field variations. Often, the pH test is used as a guide for the range of lime percentages used in laboratory strength testing.

Amount of Water

The National Lime Association has long recommended taking the moisture content up to 1-2 percent above optimum and then using manipulation and curing to drop the moisture content back to optimum for compaction.

Hydrated lime (calcium hydroxide) is required for the soil stabilization reactions to occur. The three methods of obtaining hydrated lime are (1) production in commercial lime manufacturing plants, (2) production at the jobsite in large hydration tanks specifically designed for that purpose, and (3) adding dry quicklime (calcium oxide) to the soil with subsequent hydration by adding generous amounts of water. Obviously, the last method is most prone to variation because field personnel can easily underestimate the amount of water required for hydration in the soil. If addition of dry quicklime to the soil is to be accepted, special efforts should be made to insure that enough water has been added to hydrate the quicklime, enough excess water is added to take the soil moisture up 3-4 percent above optimum, and then it is allowed to drop back to optimum through curing and manipulation.

Compaction

This is the most easily controlled variable since testing procedures are well established. Generally, lime treatment makes soil much easier to reach compaction. If the contractor is unable to reach specified density, recalibration of the testing equipment is suggested because this is one of the most frequently encountered problems. One precaution to consider in achieving final grade on the site is the associated decrease in achievable maximum density in lime-treated soils because of flocculation/agglomeration of the clay particles. This leads to what is commonly referred to as "fluffing" of the soil, where the final grade after compaction may be slightly higher than expected because of the lowered compacted density, and this should be considered in initial design planning.

Curing

Experience has shown that curing is best accomplished when the compacted soil is covered with an appropriate asphalt emulsion. Curing with water has been accomplished and works acceptably provided the compacted lime-treated soil is kept sufficiently wet on a continuous basis. However, this is generally impractical and the use of asphalt emulsion is highly preferred.

High Sulfate Soils

Soils high in sulfates are a special problem that should be carefully evaluated prior to design and construction. Much research has been conducted in the last few years and has resulted in the industry adopting special procedures, part of which are outlined in Chapter 4. Generally, it is believed that current techniques can mitigate many problematic sulfate soils. However, adequate field and laboratory testing of the site soils is essential if the project is in a high sulfate area.

CHAPTER 11

CONSIDERATIONS FOR USING LIME STABILIZATION IN PAVEMENT RECYCLING

11.01 Recycling Alternatives

Recycling or the reuse of existing paving materials for pavement rehabilitation is not a new concept (Epps et al., 1980). Categorization of recycling approaches is usually based on (1) the recycling procedure used, (2) the type of paving materials to be recycled and the end products they are to produce, or (3) the structural benefit to be gained from the recycling approach (Epps et al., 1980).

Recycling alternatives have been divided by Epps et al. (1980) into three categories:

1. **Surface Recycling**—Reworking the surface of a pavement to a depth of less than about 25.4-mm (1-inch) by heater-planer, heater-scarifier, hot-milling, cold-planing or cold-milling devices. This operation is a continuous, single-pass, multistep process that may involve use of new materials, including aggregate, modifiers or mixtures. This process is described in depth by Button, Little and Estakhri (1993).
2. **In-Place Surface and Base Recycling**—in-place pulverization to a depth greater than about 25.4-mm (1-inch), followed by reshaping and compaction. This operation may be performed with or without the use of a stabilizer.
3. **Central-Plant Recycling**—scarification of the pavement material, removal of the pavement from the roadway prior to or after pulverization, processing of material with or without the addition of a stabilizer or modifier and laydown and compaction to desired grade.

Recycling alternatives 2 and 3 may call for the use of stabilizers for the existing materials. The selection of lime in these applications is related to the benefits which can be gained by using lime to:

1. Upgrade existing base or subbase materials through modification of fines (i.e., plasticity reduction and/or strength enhancement),
2. Enhance the bond between existing aggregate materials and other stabilizers due to the surface interactive effects of lime and many aggregates and
3. Reduce stripping potential and moisture damage potential of asphalt-aggregate mixtures due to the ability of lime to reduce age harding at aggregate-asphalt interfaces and the ability of lime to promote improved, moisture resistant bonding between the asphalt binder and the aggregate interface.

In-Place Recycling

A major advantage of in-place recycling is the ability to significantly improve the load-carrying capability of the pavement without changes in the horizontal and vertical geometry of the roadway. In-place recycling also offers the ability to treat almost all types of pavement distress in asphalt-surfaced roadways, to reduce or eliminate reflection cracking, to reduce frost-susceptibility of the recycled material and to improve skid resistance and the ride quality of the roadway (Epps et al., 1980).

The major disadvantage of in-place recycling is that quality control is not as good as for central-plant operations.

Central Plant Recycling

Central plant recycling of base and surface pavement layers has been practiced for many years. The central plant process usually refers to placing recycled asphalt pavement (RAP) back through an asphalt paving plant modified to accommodate the RAP and to add the modifiers or recycling agents necessary to upgrade the quality of the RAP.

Central plant recycling can also refer to recycling of materials for base or subbase construction through a central plant instead of recycling these materials in-place. This central plant operation for base materials is normally quite simple and efficient. The plant requires a storage hopper for new aggregate (if any) needed to alter gradations of the recycled aggregate to meet specified levels, a storage hopper for the required stabilizers to be added to the recycled base, a source of water to provide appropriate moisture for mixing and compaction and mixing equipment to blend the recycled material, new aggregate and water. This mixing equipment can be as simple as conveyor belt plows.

Although central plant base recycling operations allow for better quality control than most in-place operations, modern effective and efficient mixing equipment has made in-place base recycling more attractive and more widely used than central plant base recycling.

11.02 Analysis and Design Steps Required for In-Place Recycling

Equipment and Methods

The types of equipment used for in-place recycling are very similar to those used for in-place stabilization with lime as discussed in Chapter 9. The only specialized equipment is that necessary to properly pulverize bound material prior to restabilization.

Specially designed pulverizers, hammer mills, or cold milling machines have been developed for these purposes.

The basic sequence of events for in-place recycling with the addition of a stabilizer is as follows:

1. Rip and break the existing asphalt pavement with a dozer with a ripper; with a pulverizer, if the surface layer is thin enough; or simply with a maintainer with scarifier teeth;
2. Pulverize the existing pavement including the asphalt surface and the existing base to meet specifications;
3. Add the appropriate amount of lime or lime and fly ash or other stabilizers to the pulverized paving material;
4. Mix the pulverized, recycled material with the stabilizer in accordance with specifications and be careful to add sufficient moisture for the development of basic stabilization reactions including pozzolanic reaction;
5. Spread the upgraded base to the appropriate thickness;
6. Compact, seal and cure in accordance with specification and
7. Apply wearing course.

Application of In-Place Recycling Techniques

The application of in-place recycling offers several advantages (Epps et al., 1980):

1. Equipment requirements are minimal and, besides ripping or milling equipment required for some of the thicker asphalt concrete surfaces, is the same equipment required for in-place lime stabilization;
2. In-place recycling affords the opportunity to correct structural and material problems quickly and, therefore, without prolonged disruption of traffic;
3. In some cases the residual asphalt acts as an excellent binder to help make the recycled base more resistant to water and less frost susceptible and
4. The addition of a stabilizer, such as lime or lime and fly ash, may upgrade the recycled base by reducing swell potential where active clays are present, reduce freeze-thaw potential and/or increase the load-carrying capacity of the pavement structure.

Construction Procedures

The basic construction steps of recycling with lime are identical to those discussed in Chapter 9, once the existing pavement has been ripped and pulverized and is ready for the application of lime. The remixing, compaction, curing and sealing steps are identical.

Mixture Design

The same basic steps of mixture design as discussed in Chapter 5 apply for the mixture design of the recycled material.

Structural Design

The same basic steps and considerations of structural design and analysis as discussed in Chapters 6 and 7 are applicable to lime stabilized recycled materials.

Energy and Cost Considerations

Costs associated with in-place recycling are discussed by Epps et al. (1980). Although the costs used by Epps et al. are outdated, the approach to considering energy and costs is valid as is the life cycle cost approach presented in Chapter 8.

Construction Specifications and Quality Control

Construction specifications and quality control specifications presented in Chapter 10 are valid for recycled pavements.

11.03 Case History of the Use of Lime in In-Place Recycling

Recycling of City Streets in Waco, Texas

Overview of Project

The city of Waco, Texas, like most cities suffered from badly deteriorating residential streets. Their policy had been to reconstruct these streets by removing the worn-out streets and rebuilding them with new material, a very costly process.

Planning and Construction Process

Waco city planners made a detailed study of the 227.5 km (140 miles) of residential streets to determine which would be good candidates for reconstruction. Selection was based on the availability of proper quality and quantity of existing gravel base material, presence of sewer and water lines in satisfactory condition and absence of drainage problems (Reinhardt, 1992).

After street selection, the next step was to adjust the manholes and water valves to a depth of 305 to 457-mm (12 to 18-inches) below existing grade. A grader-scarifier then scarified the old asphalt surface and base course to a depth of 204 to 254-mm (8 to 10-

inches), followed by pulverization and pre-mixing with a Bomag stabilizer. The material was then inspected to determine which soil stabilizer to use—hydrated lime, Portland cement or asphalt emulsion. Because of the abundance of plastic clay present in the base material, lime was selected for nearly all reconstruction. During the initial pre-mixing operation, excess material was removed and stockpiled for future use. Following premixing, the material was shaped and recompacted lightly to permit traffic to use the street during the pre-stabilization period (Reinhardt, 1992).

The stabilization procedure included scarifying, adding lime in slurry form, mixing, compaction, shaping and curing. The slurry lime was prepared in a Portabatch lime slaker at about 32 percent solids. The lime was delivered and spread by slurry trucks handling 9.1 to 10.9 metric tons (10 to 12 tons) of lime solids. The lime was spread to the desired rate of 13.3 kg/m^2 (25 lb./sq. yd.) for a depth of 204-mm (8-inches) (approximately a 4 percent application) (Reinhardt, 1992).

Once the proper gradation and water content were attained, the lime-stabilized gravel base was shaped and compacted with sheepsfoot, pneumatic and flat wheel rollers in succession until proper density was achieved. A prime coat of MS-1 emulsion and water mixed at a rate of 3.8 liters (1 gallon) of emulsion to 10 liters (2.5 gallons) of water was applied at 1 liter/m^3 (0.25 gal./sq. yd.). The stabilized base was cured for 1 to 2 weeks prior to paving with only light traffic permitted during the curing period (Reinhardt, 1992).

Cost Savings

Waco city officials are very satisfied with the performance of the recycled pavements. They believe that the added stiffness and strength provided by the lime-stabilized recycled base allows them to produce reconstructed pavements that can serve well over an expected life of 30-years and which are comparable or better than newly constructed pavements (Reinhardt, 1992).

The cost per linear meter of completed recycled street was estimated to be $51.38 ($15.81 per linear foot) as compared with $162.50 ($50 per linear foot) for undercutting and wasting old material and reconstructing with new material (Reinhardt, 1992).

Project Success as Determined From Visual and Serviceability Data

The recycled pavements have provided excellent serviceability. The most noticeable distress of the streets prior to recycling and of other old residential streets which are in need of recycling is the substantial level of deep layer rutting due to the soft and low-stability gravel base contaminated with plastic clay. The recycled street with lime application to stabilize the existing gravel, clay and pulverized asphalt is much more rut resistant and has shown no signs of rutting distress.

Yolo California Recycling Project

In 1976 through 1979 Little (1979) made an extensive structural evaluation of recycled paving materials. One particularly impressive project was on Highway 45 near Yolo, California. In this project an aggregate base which was contaminated with clay fines was recycled and lime was added in the recycling process. The original pavement consisted of 102-mm (4-inches) of asphalt concrete surface and approximately 254 to 305-mm (10 to 12-inches) of aggregate base course with plastic clay fines. The recycled pavement consisted of 254-mm (10-inches) of the existing base course restabilized with hydrated lime. This recycled base was covered with 51-mm (2-inches) of new hot mix asphalt concrete.

Non-destructive testing of the pavement sections with the Dynaflect demonstrated that the recycled base vastly improved the load carrying potential of the recycled pavement as compared to the pavement section prior to recycling. Little (1979) states that, based on a dual parametric analysis of the maximum surface deflection and the shape of the deflection basin, the recycling operation improved the structural capacity of the pavement by approximately 400 percent.

11.04 References

Epps, J. A., Terrell, R. L., and Little, D. N., (1980). "Guidelines for Recycling Pavement Materials," National Cooperative Highway Research Program Report 224.

Button, J. W., Estakhri, C., and Little, D. N., (1993). "Hot In-Place Recycling," National Cooperative Highway Research Program Synthesis Report 1–10b.

Little, D. N., (1979). "Structural Properties of Recycled Pavement Mixtures," Ph.D. Dissertation, Texas A&M University.

Reinhardt, K., (1992). "Street Rehabilitation Program Uses Hydrated Lime for Stabilization," Public Works, March.

CHAPTER 12

LIME SLURRY PRESSURE INJECTION

12.01 Introduction

Lime slurry pressure injection (LPSI) has proven to be effective for stabilizing expansive foundation soils in construction. This process is treated in detail in NLA Bulletin 331. The LPSI process involves injecting hydrated lime slurry under pressure to depths of 1 to 3.1-m (3 to 10 feet), and occasionally 12.3-m (40 feet) or more. The slurry moves through the soil by following paths of least resistance and is forced laterally and vertically into cracks, fissures, root holes, etc. in the soil structure, (Figure 12.1). The lime reacts with the clay soil at the interface of the seam to increase the soil strength and reduce moisture movement by physical and chemical changes that occur due to soil-lime reactions. The net effect is not only strength increase but also a reduction in moisture movement and the result of the damaging effects of moisture fluctuation in the soil.

The nature of expansive soils is to shrink in dry periods and expand in wet periods. This shrink-swell reaction causes severe damage to structures built over the soil such as foundations and roadways. In dry periods the soils develop tension cracks due to lack of moisture, and in time of high rainfall they expand and swell. This volume change can be as great as 30 percent in linear swell or shrinkage. This cyclic wet-dry action can also induce the type of fractured and fissured structure that is conducive to successful LSPI.

In recent years fly ash has enjoyed increased use in lime injection stabilization. Generally lime-fly ash slurry results in a more pronounced increase in the bearing strength of silty and sandy soils than soils deficient in reactive minerals where lime alone may not be effective.

Pressure injection increases the strength of embankments by adding tensile reinforcing strength, mending existing cracks and causing peak strength of the embankment fill and the peak strength of the foundation subsoil to be mobilized simultaneously, thus reducing progressive failure effects.

12.02 Lime Slurry Injection Process

The LSPI process consists of pumping a slurry of hydrated lime and water containing 22-36 percent lime solids into the subgrade. Injections are made vertically into the soil with holes typically spaced on a 1.54-m (5-foot) grid pattern. Initial injections are often

FIGURE 12.1. LSPI MOVES THROUGH THE SOIL BY FOLLOWING THE PATHS OF LEAST RESISTANCE (AFTER NLA BULLETIN 331).

followed by secondary or even tertiary injections, spaced diagonally between the previous injections. Depth of injection will vary based on specific job site conditions.

Slurry pressure and flow are obtained from a suitable pump, which is mounted on a slurry mixing tank which is equipped with a mechanical agitator and is capable of bulk mixing a 18.1 to 22.7 metric ton (20–25 ton) truckload of hydrated lime with 60.6 m^3 (16,000 gallons) of water. The resultant lime slurry is pumped at a pressure of 345 to 1,379 kPa (50 to 200 psi) through a high pressure hose to the injection rig. Slurry is injected at frequent depth intervals to refusal or in a slow continuous path until a specified quantity is injected.

The amount of lime required for LSPI treatment can vary considerably, depending on soil properties, injection depth, permeability of the soil mass and degree of stability required. A typical value of slurry required is in the range of 0.27 to 0.38 Kg/m^3 (0.6 pounds to 0.85 pounds per cubic foot) for a single injection and about 0.45 to 0.68 Kg/m^3 (1.00 pound to 1.50 pounds per cubic foot) for a double injection. Figure 12.2 illustrates a typical grid pattern of LSPI.

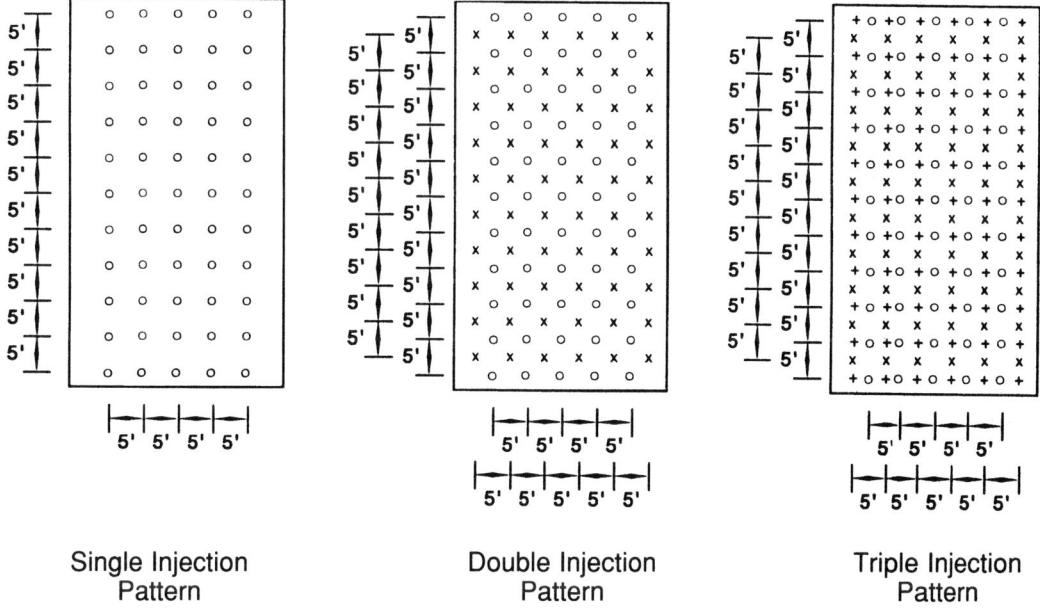

FIGURE 12.2. THE GRID SYSTEM FOR LSPI MAY VARY DEPENDING ON THE NATURE OF THE PROJECT (AFTER NLA BULLETIN 331).

12.03 Injection Materials

Besides lime, a surfactant and sometimes fly ash are used in LPSI. The surfactant is a wetting agent and is commonly used in the lime slurry to reduce surface tension and promote better slurry penetration into the soil mass. A non-toxic surfactant is commonly used and is added at a rate of approximately one part surfactant to 3,500 parts water.

Fly ash is a pozzolan and is a by-product of the burning of coal. The reaction between the available silica and alumina in fly ash and the calcium in the lime and water added for construction purposes can result in significant pozzolanic reactions and high strengths in soils not normally reactive with lime alone.

12.04 LSPI Mechanisms

The basic reaction mechanisms involved in soil-lime mixtures are discussed in some detail in Chapter 4. The mechanisms of LSPI depend on the basic reactions discussed in Chapter 4. However, since the LSPI process does not involve the intimate mixing that is involved in the standard stabilization process, the mechanisms of stabilization of the soil mass are somewhat different from, albeit dependent upon, the basic reactions of cation exchange and pozzolanic reaction. When used in new construction over plastic soils, LSPI actually wets and pre-swells the soil and then tends to retain the high moisture level with its network of lime seams that form moisture resistant barriers.

When expansive clay soils are injected with lime slurry, a number of changes result which together improve the engineering characteristics of the soil mass. These include: stabilizing effect of lime seams, preswelling, translocation, supernate penetration and the stabilizing effect of the lime/fly ash seams.

Stabilizing Effect of Lime Seams

The lime seams stabilize the moisture content of the treated soil mass by forming a barrier which impedes the movement of capillary as well as seasonal moisture through the soil. By encapsulating large volumes of clay, the volume change potential of the soil is greatly reduced. A second benefit of the lime seams is the pozzolanic strength reaction at the interface of the stabilization seams and the adjoining clay soil.

Preswelling

The injection process increases the moisture content of the soil by approximately 2–3 percentage points. This relatively high moisture level can be thought of as preswelling. However, if preswelling of the clay soil were done to this level of moisture content without the addition of lime, the soil could result in an unacceptable loss of bearing capacity and thus instability.

The addition of lime or lime and fly ash in the injection process not only allows preswelling of the clay without loss of strength but also helps to maintain a stable moisture content as the lime stabilized soil is less likely to loose moisture than the unstabilized soil.

Translocation

After the stabilization seams are formed, some of the lime translocates or migrates and modifies the soil adjacent to the seams and can result in a gradual strength gain throughout the soil mass. Blacklock (1982) states that "the LSPI method of soil stabilization and reinforcement results in the formation of a network of thin sheets and seams. These lime slurry sheets and seams react with the adjacent soil to form strong, relatively impervious tensile membranes locked into the soil mass. The effect of these membranes is to control the movement of moisture and to reinforce and confine the segmented portions of the soil mass."

Supernate Penetration

Lime supernate (slurry water) has a pH of from approximately 11.0 to 12.4. This causes the supernate of the lime slurry to be drawn into the soil between the seams either by soil suction or diffusion mechanisms. This makes it possible for the soil among the seams to be impregnated with slurry water which in turn can force the cation exchange reaction. This concept of supernate penetration is supported by the work of Stocker (1972) and Petry (1980).

12.05 Candidate Soils for LSPI

Compression and shear strength tests should be used to evaluate sites with low strength soils; swell tests should be used to evaluate sites with potential settlement problems. Standard soil classification tests and other index tests such as Atterberg limit tests should not be used as substitutes for strength and volumetric tests.

The testing procedures should simulate LSPI field conditions as closely as possible. This involves treating the soil samples with lime slurry to form a glaze or seam, then curing and testing. The results of the tests on the glaze or seam lime stabilized test samples are then compared to control samples. The amount of dry lime solids used in LPSI compatibility testing is usually one percent of the soil dry weight. This has been determined to be the maximum amount of lime injected during a single stage LSPI spaced on 1.54-m (5-foot) centers on a diagonal offset pattern.

Lime glaze or seam stabilized samples can be used in swell, consolidation and compression testing. This testing method was developed by Blacklock (1977).

The purpose of the testing program is to determine whether lime slurry or lime and fly ash slurry will improve the candidate soil site and to guide in preparing appropriate specifications. However, it is not possible to obtain exact correlations between laboratory test results and the precise degree of success obtainable in the field. The comparative results of testing on the stabilized samples and the control samples does give a relative comparison.

The engineering testing program suggested by Bulletin 331 of the National Lime Association, "Lime Slurry Pressure Injection Bulletin," suggests the use of several evaluation tests to determine whether or not LSPI and lime and fly ash pressure injection are appropriate. The suscint explanations presented below are provided only as a brief definition of the test. Detailed descriptions are offered in Bulletin 331.

Glaze Stabilized Compression Tests

Bulletin 331 discusses the use of the glazed stabilized compression test as illustrated in Figure 12.3. The purpose of this test is to determine additional strength afforded to

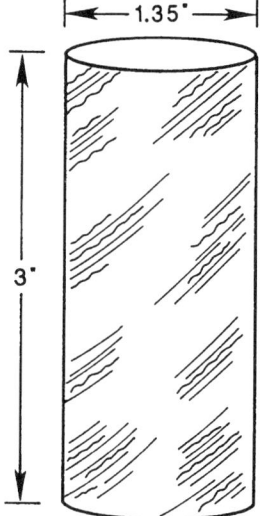

FIGURE 12.3. THE GLAZED STABILIZED COMPRESSION TEST PROVIDES INFORMATION AS TO THE STRENGTH GAINED BY GLAZED COATING OF THE SOIL (AFTER NLA BULLETIN 331).
1 in = 25.4 mm

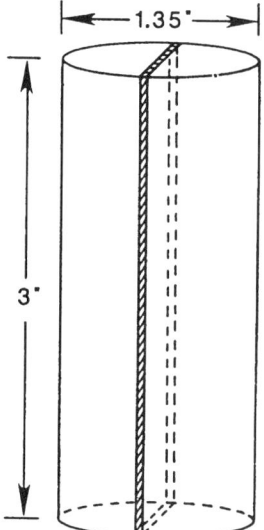

FIGURE 12.4. THE STRAIGHT SEAM COMPRESSION TEST IS DESIGNED TO EVALUATE THE REINFORCEMENT GAINED THROUGH VERTICAL LSPI SEAMS (AFTER NLA BULLETIN 331).
1 in = 25.4 mm

the sample by the reinforcing action of the glaze stabilized coating. The glaze-stabilized samples are compared to unstabilized samples prepared and cured in an identical manner to the treated samples.

Seam Stabilized Compression Test

Two types of seam tests are performed: straight split (Figure 12.4) and angle seam (Figure 12.5). The straight seam sample is designed for evaluation of the compression strength reinforcement component of the stabilized seam and the angle seam sample is designated for evaluation of the stabilized seam (shear) strength. Normally, the contributions of both compression and shear will be utilized in repairing cracks in existing embankment failures. These samples can be prepared from undisturbed soil samples, but experience indicates a preference for remolded samples. These can also be glazed coated to allow evaluation of combinations of shear, tension and compressive strength reinforcement (NLA Bulletin 331).

Glaze Stabilized Consolidation Test

The purpose of this test is to evaluate the reduction in settlement potential provided by injection stabilization of natural embankment soils. This sample is prepared by cutting undisturbed samples and then applying a glaze stabilization coating to both the top and bottom surface of the samples (NLA Bulletin 331).

Seam Stabilized Swell Test

The purpose of this test is to evaluate the swell reduction function of lime and lime and fly ash seams. A typical sample is prepared by remolding soil and placing a slurry grout seam in the center (NLA Bulletin 331).

Material Test

Besides the six soil stabilization tests discussed, it is necessary to test all source materials. Since fly ash is a variable material whose strength-related properties are dependent on the source of the ash, conditions of pulverization and burning of the coal and storage of the ash, it is best to evaluate the ash and combinations of ash and lime separately by performing cube or compression cylinder tests. The tests should evaluate time, temperature and strength variables for different mixing times, different mix ratios and different material manufacturers or sources (NLA Bulletin 331).

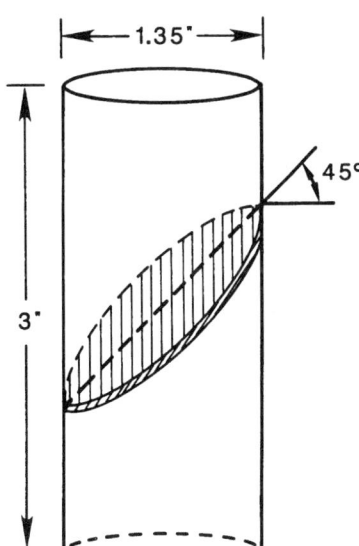

FIGURE 12.5. THE ANGLE SEAM COMPRESSION TEST IS DESIGNED TO EVALUATE THE REINFORCEMENT GAINED THROUGH LSPI SEAMS WHICH OCCUR AT AN ANGLE (AFTER NLA BULLETIN 331).

Field Pump Tests

A field pump test during the design stage may be necessary to determine the slurry volume placed with a single or double injection. It may be desirable to dig a trench to observe slurry flow in the trench side walls, especially if there is a question about available fissures and openings in the soil mass to accept the slurry. This type of data can be obtained from Shelby tube samples and radiographing them to determine fissures, cracks and other anomalies that will accept the slurry.

Surcharge Tests

For certain sites where consolidation is a problem, it may be necessary to inject a test pad and then surcharge the pad as well as a control section and monitor the results.

The Decision Process

There is no simple method of obtaining a yes-or-no answer for all possible LSPI sites. The tests and other evaluations outlined in Bulletin 331 and summarized here will provide meaningful engineering data to aid in the decision process.

The ultimate question is: "Will the injection of lime slurry improve the soil mass, and if so, how much?"

12.06 Safety Precautions

Safety and general precautions for using LSPI are discussed in Bulletin 331.

12.07 Conclusions

Although a general perception is that a soil mass should be dry and highly fractured and fissured to accommodate the flow of LSPI for stabilization, experience has shown that nearly all expansive clays can be injected with LSPI since there appears always to be desiccation cracks and fissures through which the slurry can flow. The highly plastic and highly expansive Yazoo clay of Mississippi has been successfully injected (NLA Bulletin 331).

Bulletin 331 states that even when clays are wet, the fissures are still present due to the non-elastic nature of soil. However, when a "tighter", more plastic and less fissured clay is encountered, it is usually necessary to use closer spacings and more than one injection pass.

12.08 References

Blacklock, J. R., and Lawson, C. H., (1977). "Handbook for Railroad Track Stabilization Using Lime Slurry Pressure Injection," Federal Railroad Administration Report FRA/ORD-77/30.

Blacklock, J. R., Joshi, R. C., and Wright, P. J., (1982). "Pressure Injection Grouting of Landfills Using Lime and Fly Ash," *Proceedings, ASCE Specialty Conference, Grouting in Geotechnical Engineering*, New Orleans.

National Lime Association. *Lime Slurry Pressure Injection Bulletin*, Bulletin 331.

Petry, T. M. and Armstrong, J. C., (1980). "Relationships and Variations of Properties of Clay," Proceedings of the Fourth International Conference on Expansive Clays, Vol. 1, ASCE, Denver, Colorado.

Sterber, P. T., (1972). "Diffusion and Diffuse Cementation in Lime and Cement Stabilized Clayey Soils," Special Report No. 8, Australian Road Research Board.